NEUROMETHODS

Series Editor
Wolfgang Walz
University of Saskatchewan
Saskatoon, SK, Canada

For further volumes:
http://www.springer.com/series/7657

Neuromethods publishes cutting-edge methods and protocols in all areas of neuroscience as well as translational neurological and mental research. Each volume in the series offers tested laboratory protocols, step-by-step methods for reproducible lab experiments and addresses methodological controversies and pitfalls in order to aid neuroscientists in experimentation. *Neuromethods* focuses on traditional and emerging topics with wide-ranging implications to brain function, such as electrophysiology, neuroimaging, behavioral analysis, genomics, neurodegeneration, translational research and clinical trials. *Neuromethods* provides investigators and trainees with highly useful compendiums of key strategies and approaches for successful research in animal and human brain function including translational "bench to bedside" approaches to mental and neurological diseases.

Translational Research Methods for Alcohol Use Disorders

Edited by

Melissa A. Cyders

Department of Psychology, Indiana University – Purdue University, Indianapolis, IN, USA

Humana Press

Editor
Melissa A. Cyders
Department of Psychology
Indiana University – Purdue University
Indianapolis, IN, USA

ISSN 0893-2336 ISSN 1940-6045 (electronic)
Neuromethods
ISBN 978-1-0716-3266-6 ISBN 978-1-0716-3267-3 (eBook)
https://doi.org/10.1007/978-1-0716-3267-3

This Humana imprint is published by the registered company Springer Science+Business Media, LLC, part of Springer Nature.
The registered company address is: 1 New York Plaza, New York, NY 10004, U.S.A.

Preface to the Series

Experimental life sciences have two basic foundations: concepts and tools. The Neuromethods series focuses on the tools and techniques unique to the investigation of the nervous system and excitable cells. It will not, however, shortchange the concept side of things as care has been taken to integrate these tools within the context of the concepts and questions under investigation. In this way, the series is unique in that it not only collects protocols but also includes theoretical background information and critiques which led to the methods and their development. Thus it gives the reader a better understanding of the origin of the techniques and their potential future development. The Neuromethods publishing program strikes a balance between recent and exciting developments like those concerning new animal models of disease, imaging, in vivo methods, and more established techniques, including, for example, immunocytochemistry and electrophysiological technologies. New trainees in neurosciences still need a sound footing in these older methods in order to apply a critical approach to their results.

Under the guidance of its founders, Alan Boulton and Glen Baker, the Neuromethods series has been a success since its first volume published through Humana Press in 1985. The series continues to flourish through many changes over the years. It is now published under the umbrella of Springer Protocols. While methods involving brain research have changed a lot since the series started, the publishing environment and technology have changed even more radically. Neuromethods has the distinct layout and style of the Springer Protocols program, designed specifically for readability and ease of reference in a laboratory setting.

The careful application of methods is potentially the most important step in the process of scientific inquiry. In the past, new methodologies led the way in developing new disciplines in the biological and medical sciences. For example, physiology emerged out of anatomy in the nineteenth century by harnessing new methods based on the newly discovered phenomenon of electricity. Nowadays, the relationships between disciplines and methods are more complex. Methods are now widely shared between disciplines and research areas. New developments in electronic publishing make it possible for scientists that encounter new methods to quickly find sources of information electronically. The design of individual volumes and chapters in this series takes this new access technology into account. Springer Protocols makes it possible to download single protocols separately. In addition, Springer makes its print-on-demand technology available globally. A print copy can therefore be acquired quickly and for a competitive price anywhere in the world.

Saskatoon, SK, Canada ***Wolfgang Walz***

Preface

Although many individuals experiment with alcohol use at some point in their lives, only a few go on to develop disorders related to alcohol. Research in alcohol use disorder (AUD) has a long history of focusing both on preclinical and clinical models from predisposition to alcohol response, to the development of prevention and intervention strategies. However, due to the complexity of methodologies utilized in this field, research in AUD is often isolated either within preclinical or clinical models. Collaboration across preclinical and clinical research of AUD would accelerate scientific progress, capitalizing on the strong experimental controls offered by preclinical work, as well as the important ecological validity afforded by clinical models. This book seeks to review existing translational research methods for AUD, highlighting cross-species complementary approaches and facilitating collaborative translational work that can advance scientific progress in this field. The target audience includes student and researchers who aim to conduct translational research in AUD.

Indianapolis, IN, USA *Melissa A. Cyders*

Contents

Contributors

PETER B. BARR • *Department of Psychiatry and Behavioral Sciences, SUNY Downstate Health Sciences University, Brooklyn, NY, USA; VA New York Harbor Healthcare System, Brooklyn, NY, USA*

RICHARD L. BELL • *Department of Psychiatry, Indiana University School of Medicine, Indianapolis, IN, USA; Stark Neurosciences Research Institute, Indiana University School of Medicine, Indianapolis, IN, USA*

BENJAMIN L. BEREY • *Center for Alcohol and Addiction Studies, School of Public Health, Brown University, Providence, RI, USA*

KRISTY BIXLER • *Department of Psychiatry, Indiana University School of Medicine, Indianapolis, IN, USA*

TIFFANY M. BRAVO • *Department of Psychiatry, Indiana University School of Medicine, Indianapolis, IN, USA*

ELIZABETH M. BURNETTE • *Department of Psychology, University of California, Los Angeles, CA, USA*

MCKENZIE R. COX • *Department of Psychiatry, Indiana University School of Medicine, Indianapolis, IN, USA*

KATHLEEN I. CRUM • *Department of Psychiatry, Indiana University School of Medicine, Indianapolis, IN, USA*

MELISSA A. CYDERS • *Department of Psychology, Indiana University Purdue University Indianapolis, Indianapolis, IN, USA*

ERICA N. GRODIN • *Department of Psychology, University of California, Los Angeles, CA, USA*

MEREDITH HALCOMB • *Department of Radiology and Imaging Sciences, Indiana University School of Medicine, Indianapolis, IN, USA*

SHEKETHA R. HAUSER • *Department of Psychiatry, Indiana University School of Medicine, Indianapolis, IN, USA; Stark Neurosciences Research Institute, Indiana University School of Medicine, Indianapolis, IN, USA*

JAMES M. HAYS • *Department of Psychiatry, Indiana University School of Medicine, Indianapolis, IN, USA*

FREDERIC W. HOPF • *Department of Psychiatry, Indiana University School of Medicine, Indianapolis, IN, USA*

LESLIE A. HULVERSHORN • *Department of Psychiatry, Indiana University School of Medicine, Indianapolis, IN, USA*

J. DAVID JENTSCH • *Department of Psychology, Binghamton University, Binghamton, NY, USA*

ANN E. K. KOSOBUD • *Department of Neurology, Indiana University School of Medicine, Indianapolis, IN, USA*

ROBERT F. LEEMAN • *Department of Health Sciences, Bouve College of Health Sciences, Northeastern University, Boston, MA, USA*

JACQUELYN L. MEYERS • *Department of Psychiatry and Behavioral Sciences, SUNY Downstate Health Sciences University, Brooklyn, NY, USA; VA New York Harbor Healthcare System, Brooklyn, NY, USA*

STEVEN J. NIETO • *Department of Psychology, University of California, Los Angeles, CA, USA*
SEAN O'CONNOR • *Department of Psychiatry, Indiana University School of Medicine, Indianapolis, IN, USA*
MARTIN H. PLAWECKI • *Department of Psychiatry, Indiana University School of Medicine, Indianapolis, IN, USA*
VIJAY A. RAMCHANDANI • *Human Psychopharmacology Laboratory, Division of Intramural Clinical and Biological Research, National Institute on Alcohol Abuse and Alcoholism, Bethesda, MD, USA*
LARA A. RAY • *Department of Psychology, University of California, Los Angeles, CA, USA; Department of Psychiatry and Biobehavioral Sciences, University of California, Los Angeles, CA, USA*
ZACHARY A. RODD • *Department of Psychiatry, Indiana University School of Medicine, Indianapolis, IN, USA; Stark Neurosciences Research Institute, Indiana University School of Medicine, Indianapolis, IN, USA*
THATIANE DE OLIVEIRA SERGIO • *Department of Psychiatry, Indiana University School of Medicine, Indianapolis, IN, USA*
BETHANY L. STANGL • *Human Psychopharmacology Laboratory, Division of Intramural Clinical and Biological Research, National Institute on Alcohol Abuse and Alcoholism, Bethesda, MD, USA*
PHILLIP A. STARSKI • *Department of Psychiatry, Indiana University School of Medicine, Indianapolis, IN, USA*
JESSICA WEAFER • *Department of Psychology, College of Arts and Sciences, University of Kentucky, Lexington, KY, USA*
BAILEE R. WILSON • *Department of Health Sciences, Bouve College of Health Sciences, Northeastern University, Boston, MA, USA*
KARMEN K. YODER • *Center for Neuroimaging, Department of Radiology and Imaging Sciences, Indiana University School of Medicine, Indianapolis, IN, USA; Stark Neurosciences Research Institute, Indiana University School of Medicine, Indianapolis, IN, USA*

Chapter 1

Genetics and Family History of Alcohol Use Disorders

Peter B. Barr and Jacquelyn L. Meyers

Abstract

Alcohol use disorders (AUD) are complex traits that are moderately heritable. A variety of approaches have been developed that allow researchers to better understand their genetic etiology. We cover several approaches for incorporating information on family history and genetic risk into research on AUD. Latent genetic designs leverage information from biological relatives. The classic twin design helps to understand population variability in AUD, while other approaches using family data can be applied to better understand individual risk. Measured genotypic approaches, such as genome-wide association studies (GWAS), use information from the entire genome to identify individual variants associated with risk for AUD. GWAS results can then be transformed into aggregate measures of individual risk in the form of polygenic scores (PGS) in non-overlapping samples. We discuss specific approaches to these methods, various considerations for each, and the limitations with currently available tools.

Key words Genome-wide association study, Polygenic score, Gene-environment interaction, Genetic epidemiology

1 Introduction

Alcohol use disorder (AUD) is a condition characterized by an impaired ability to stop or control alcohol use despite adverse social, occupational, or health consequences. AUD are associated with substantial cost to society, affected individuals, and their families [1–4]. Importantly, AUD are generally missed in routine medical care [5]. Even when they are identified, few with AUD go on to receive treatment [6]. Given the substantial human and economic costs of AUD and how often those with problems miss the opportunity to receive necessary treatment, developing methods of identifying persons at risk for AUD is vital.

The development of alcohol use disorders is complex and multifactorial, with risk factors ranging from broader social conditions to individual differences [7–9]. AUD are moderately heritable, with ~50% of the variance being attributable to genetic variation between individuals [10]. While related, alcohol *use* and

Melissa A. Cyders (ed.), *Translational Research Methods for Alcohol Use Disorders*, Neuromethods, vol. 201,
https://doi.org/10.1007/978-1-0716-3267-3_1,

AUD have only partial genetic overlap, suggesting use and problems are not interchangeable phenotypes [11, 12]. AUD shows strong overlap with other externalizing (e.g., attention-deficit/hyperactivity disorder, other substance use disorders) [13–16], internalizing (e.g., major depressive disorder, anxiety disorders) [17, 18], and psychotic disorders (e.g., bipolar disorder, schizophrenia) [19, 20]. Additionally, genetic influences on alcohol use outcomes change over time, with externalizing liability being more important in adolescence and alcohol-specific risk becoming more important as individuals age [21, 22]. Using information that allows us to understand *how* and *why* problems develop may be of use to practitioners. In this chapter, we cover various methods for exploring the genetics of alcohol use disorders. It is important to note that social and environmental influences are also integral in the development of AUD and the focus here on genetics is not meant to undermine the importance of these other phenomena.

2 Methods

2.1 Latent Genetic Designs

Before the sequencing of the human genome, methods for inferring genetic risk relied on using known patterns of genetic similarity between biologically related individuals, which we broadly refer to as twin and family studies here. Twin studies use known patterns of relatedness between monozygotic (MZ, or identical twins) and dizygotic (DZ, or fraternal twins) to infer the degree to which variation in a trait is due to additive genetic (A), shared environmental (C), and unique environmental ([E], which also includes measurement error). Decomposing the variance is possible because MZ twins share ~100% of their genetic variants (with the exception of somatic mutations), while DZ twins share 50% of their variants, on average. Both MZ and DZ pairs share a common environment, at least during early life. Therefore, we expect the within-pair correlation in MZ twins (A + C) to be twice that of DZ twins (0.5*A + C), if we assume genetic differences alone contribute to variation in that trait. Shared environmental variance is captured by the degree that the DZ twin correlation is greater than half of the MZ correlation. The degree to which each correlation is from different unity (1 − A + C) tells us the variance explained by environment not shared in common between twins, as well as measurement error. These models have many assumptions, including only additive effects, random mating among parents, and no gene-environment interactions (GxE). It should be noted that there is evidence for assortative mating (partners that are both affected with a given trait or disease) and GxE (i.e., traumatic stress, romantic relationships, parenting, peer alcohol use) in AUD [23, 24].

While twin and family models provide valuable information about the heritability of a given trait within a *population* at a specific point in time, they do not tell us anything about *individual* risk for a given disease or disorder. There are methods that leverage information from close relatives to provide estimates of individual risk. The simplest approach is a binary indicator of whether or not an individual has a family history of AUD (FH +). Family history measures can be extended to quantitative measures of density [25, 26] which use information on the degrees of relatedness for different of biological relatives, weighted by whether or not they have AUD diagnosis/symptoms. Scores are based on degree of relatedness (e.g., parents/siblings = 0.5, grandparents/aunts/uncles/half-siblings = 0.25, cousins = 0.125) and affectedness (unaffected relatives = 0), which are averaged across the number of relatives for a measure of family history density (FHD).

RIDIT (relative-to-an-identified-distribution) [27] scores are another approach [21, 22, 28]. RIDIT scores are deviations from the expected mean of 0.5 under an assumed uniform distribution and are calculated using the proportion of individuals in the preceding category (unaffected individuals) plus one-half of the proportion of individuals in the category itself (affected individuals). Given that genetic factors likely play some role in the clustering of traits within families, regression toward the population mean is expected to be less among individuals who are more genetically related. By this logic, RIDIT scores for full siblings and DZ twins are weighted half-way back toward the mean, while RIDIT scores for MZ twins are unweighted. This effectively reduces the range of the uniform distribution as the relatedness to the target person decreases (e.g., MZ twins = 0–1, DZ twins, full siblings, and parents = 0.25–0.75). These RIDIT scores can be calculated for all related persons within a pedigree and averaged to create an individual risk score for a target individual. For example, imagine AUD had a prevalence of 0.25 (25%) in our sample, including information on extended family members. Participant A reports their father had an AUD, while their mother and half-sibling did not. Participant A would therefore have a RIDIT score of = 0.4271. Whereas, Participant B who had mother, father, and a cousin with AUD would have a value of 0.5469. Table 1 below demonstrates how these scores were calculated.

Finally, family genetic risk scores (FGRS) [29–31] have been developed in extensive registry data. FGRS are derived from extended pedigrees (first- to fifth-degree relatives of a proband) and calculated in a complex, multistep process. First, age-adjusted risk for the phenotype is calculated and the trait is converted into an underlying liability distribution (*z*-score) for each member of a pedigree. An additional correction is made within first-degree relatives to account for shared environment [29–31]. For each relative within a pedigree, there are two components: the sum of the Z-

Table 1
Example of creating RIDIT scores for a trait with a 25% prevalence

	Relative	Midpoint	Relatedness	RIDIT score
Participant A	Mother	(0.125 – 0.5)	*(0.5) + 0.5	= 0.3125
	Father	(0.625 – 0.5)	*(0.5) + 0.5	= 0.5625
	Half-sibling	(0.125 – 0.5)	*(0.25) + 0.5	= 0.4063
Participant B	Mother	(0.625 – 0.5)	*(0.5) + 0.5	= 0.5625
	Father	(0.625 – 0.5)	*(0.5) + 0.5	= 0.5625
	Cousin	(0.625 – 0.5)	*(0.125) + 0.5	= 0.5156

scores and the total weighted number of relatives, which are further weighted by their degree of relatedness to the target individual. Finally, the two components are summed across all groups of relatives. A shrinkage factor is applied to this combined estimate to adjust for the number of relatives and converted to a Z-score for easier interpretation. Like the RIDIT score approach, this requires either (1) extensive knowledge about family history of close-to-distant relatives or (2) large-scale registry data that include pedigree information.

2.2 Measured Genotypic Approaches

Recent years have seen the rapid expansion of approaches using measured genotypic data rather than risk inferred from pedigrees. Early work, referred to as candidate gene approaches, focused on genes that were purported to be associated with a trait because of a hypothesized biological mechanism. This research was plagued by false positives, publication bias, and low-powered studies [32]. Meta-analyses reveal no support for most candidate gene analyses [33], with the exception of alcohol-metabolizing genes (e.g., *ADH1B*) in relation to alcohol use [34] and AUD [35]. Further, single-gene analyses do not fit with our current polygenic understanding of complex traits such as AUD, whereby phenotypes are influenced by hundreds to thousands of variants of very small effect that act in conjunction with the social environment [36].

Since the mapping of the human genome, research examining measured genetic variants that scan the entire genome has grown exponentially. Rather than focusing on a single variant with some hypothesized biological mechanism as was done in the candidate gene era, genome-wide association studies, or GWAS, test the association between a phenotype and each single nucleotide polymorphism (SNP) across all common variants in the genome (e.g., minor allele frequency, or MAF, $\geq$ 1.0%). GWASs typically require sample sizes of hundreds of thousands to millions of individuals for associations because (1) effect sizes for individual variants are extremely small and (2) GWASs produce a large number of independent tests, which give rise to false positives. The typical p-value threshold for genome-wide significance is $p < 5.0*10^{-8}$ to correct for ~1 million independent tests. Statistical power has become less of an issue, recently, due to the growth of affordable genotyping arrays, large-scale biobanks, and direct-to-consumer genetic testing companies that amass genotypic information on large samples [37].

The typical steps for performing a GWAS are as follows. First, phenotypic data is collected on individuals that provide biospecimens (e.g., saliva, whole blood), who are then genotyped using a genotyping array. Genotyping arrays differ in their coverage of the genome, with varying quality across given regions and ancestries [38]. The genotypic data are typically filtered through a variety of quality control (QC) procedures, including (but not limited to) dropping variants with genotyping errors, filtering variants/participants with large amounts of missingness, removing variants that violate Hardy-Weinberg equilibrium, and often removing rare variants [39]. These data are then used to create ancestral principal components (PCs), which allow us to empirically assign ancestries of participants, generally using information from external reference panels [40]. Current methods for genetic analyses require relatively homogenous ancestral groups, as variation in allele frequencies across geographic location (referred to as population stratification) due to processes like genetic drift can give rise to false positives. Figure 1 provides an example of the first and second PCs plotted using data from the National Longitudinal Study of Adolescent to Adult Health (Add Health) [41, 42]. These PCs demonstrate the continuous nature of human genetic variation and that differences between populations are a matter of degree rather than kind. Importantly, while ancestry assignments are correlated with socially defined racial and ethnic constructs (e.g., those who are most genetically similar to participants from reference panels sampled from present day Africa are likely to self-identify as "Black or African American"), these are distinct phenomenon. Those who fall outside of some predefined threshold meant to maintain a relatively homogenous group (typically referred to as "admixed" individuals, the gray dots in Fig. 1) are commonly excluded from analysis. In Add Health, ancestral assignment was defined as those ±1SD of the mean of PC1 and PC2 within each self-identified racial ethnic category (Non-Hispanic White, Non-Hispanic Black, Hispanic or Latino, and Asian) [43]. More rigorous approaches of ancestry assignment use external reference panels with known geographic location to project individual ancestries onto those with whom participants are most similar to in the reference panel [44, 45].

Once data have been cleaned, SNPs that have not been tagged in the genotyping array are imputed using detailed reference panels of ancestrally similar individuals, such as the 1000 Genomes [40] or the Trans-Omics for Precision Medicine (TOPMed) program [46] reference panels. After imputation, these data typically go through an additional QC, removing variants below a given MAF threshold

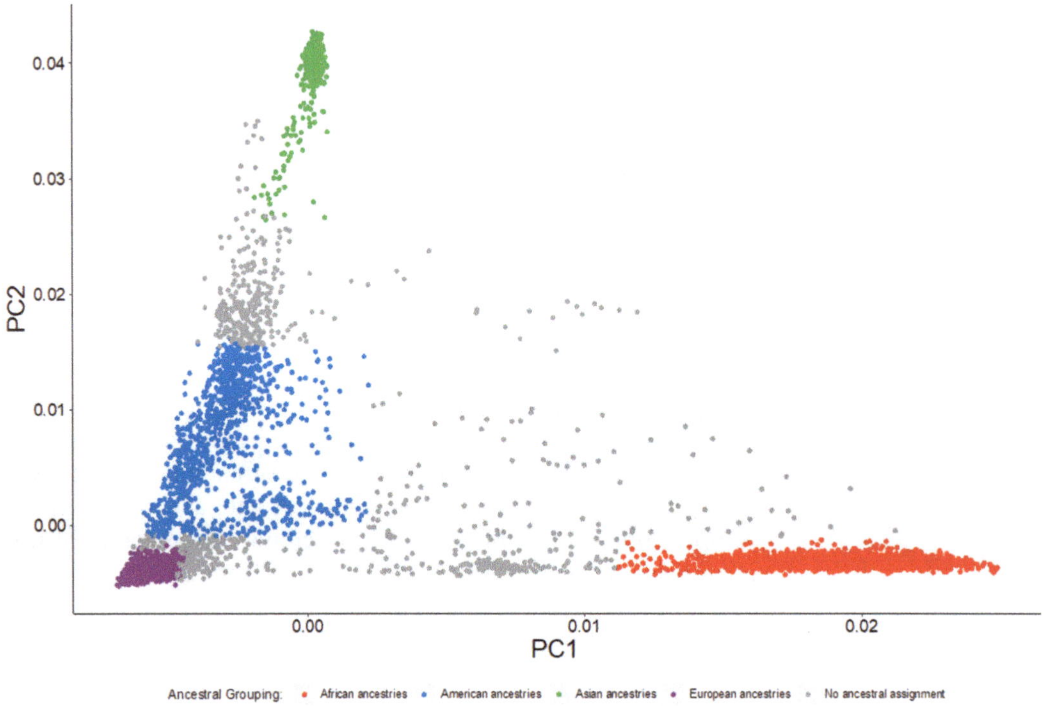

Fig. 1 Example of ancestral variability and assignment in the National Longitudinal Study of Adolescent to Adult Health (Add Health)

and variants with poor imputation quality. The last step in the process is to perform the actual tests of association between genetic variants and the phenotype of interest. GWAS are typically performed using either a linear (in the case of continuous) or logistic (in the case of binary) regression at each SNP in the genome, including covariates such as age, sex/gender, ancestral PCs (to capture additional population stratification), and any study-specific covariates (e.g., genotyping array, collection site). If one has large numbers of related individuals in their data, analyses can either (1) filter down to unrelated individuals or (2) use a mixed model where relatedness is directly modeled using a genetic relatedness matrix (GRM). Table 3 provides examples of some of the largest GWASs for alcohol-related phenotypes to date. GWAS output typically comes in the form of summary statistics for all variants included in the analysis. These files generally include, but are not limited to, columns for chromosome, position on the genome, effect (beta or odds ratio), standard error, *p*-value, coded allele, coded allele frequency, and sample size. Summary statistics can be used for a variety of downstream analyses, including bioannotations of results [47–50], genetic correlations with other traits [51–53], and causal inference [54–57].

One widely used application of GWAS summary statistics is for the creation of polygenic scores (PGS), sometimes referred to a

RSID	Beta	Coded Allele
RS1	0.025	T
RS2	0.063	A
RS3	-0.057	G

Person	SNP 1	SNP 2	SNP 3	PRS
A	AA	AT	GG	-0.089
B	AT	AT	GG	-0.026
C	AT	AA	CG	0.094

Fig. 2 Hypothetical example of calculating polygenic scores

polygenic risk scores or polygenic indices. PGS use GWAS summary statistics to create a single, aggregate measure of risk by taking the number of "effect" alleles (the coded allele in the GWAS) an individual carries and weighting those by the effect size from the GWAS (e.g., beta or odds ratio). These weighted effects are summed across the entire genome to create a single measure per person in an independent, hold-out sample (i.e., not included in the discovery GWAS). Figure 2 provides a hypothetical example of PGS construction.

While GWAS summary statistics generally include information on millions of SNPs, not all of these are included in polygenic scores. This is because SNPs that are nearby to one another on the genome (or were at some point in evolutionary history) tend to be correlated with one another, a phenomenon referred to as linkage disequilibrium, or LD. Therefore, many of the association effect sizes contain redundant information, and we must account for LD patterns when constructing polygenic scores to remove this redundant information. There are a variety of methods that have been developed for PGS construction, some of which have been listed in Table 2. Each approach has different strengths and weaknesses. Methods that correct for LD using all SNPs, such as LDPred [58] and PRS-CS [59], tend to improve predictive performance relative to classic approaches like pruning and thresholding (P + T) [60]. However, the inclusion of sub-genome-wide significant SNPs may increase the influence of population structure [61, 62], and researchers must determine the goal of PGS construction (raw prediction vs. causal explanation).

Once the method of polygenic score creation is decided, scores can be constructed in one's external validation sample. External sample participants must have no genetic overlap with initial GWAS (either themselves or related individuals), as this can bias the predictive power of PGS, upward. We present the association between a PGS from a GWAS of externalizing problems [13] and lifetime substance use disorders (SUD) in unrelated Add Health participants of European ancestries, as an example ($N = 4855$). Figure 3 (Panel a) shows how the PGS is normally distributed. This PGS shows robust association with lifetime SUD (OR = 1.438,

Table 2
Examples of methods and software for constructing PGS

Method	Description	Citation
Pruning/clumping and thresholding (P + T)	SNPs are clumped using a prespecified window and filtered at a given *p*-value threshold. These SNPs are used to create PGS, typically comparing across multiple p-value thresholds	[77]
LDPred/LDPred2	A Bayesian approach in which posterior mean effects sizes from GWAS summary statistics are calculated using a point-normal mixture distribution prior (two parameters: SNP-based heritability and the fraction of causal variants) and LD information from an external reference panel	[58, 78]
PRS-CS	A Bayesian regression framework that calculates posterior SNP effect sizes from GWAS summary statistics under continuous shrinkage priors using information from an exterior LD reference panel	[59]
SBayesR	SBayesR applies a Bayesian multiple regression model to GWAS summary statistics by combining a likelihood connecting GWAS regression coefficients and a finite mixture of normal distribution prior on the reported effects sizes. Uses a sparse LD correlation matrix from an external reference panel	[79]
Lassosum	SNP weights selected from summary statistics using an external LD reference panel and a penalized regression framework. Tuning parameters for the penalized regression can be optimized with a validation dataset or by use of pseudo-validation procedure	[80]

Table 3
Currently available GWAS of alcohol-related traits

Phenotype	Description	Sample size (N)	Ancestries	Citation
Alcohol consumption	Number of alcoholic drinks consumed per week	941,280	European	[34]
Problematic alcohol use	Meta-analysis of alcohol dependence, AUD, and the problem subscale of the AUDIT (AUDIT-P)	435,563	European	[8]
Alcohol dependence	DSM-IV alcohol dependence	52,848	European and African	[12]
Alcohol use disorders	AUD diagnosis	274,424	European, African, and Latin American	[35]
Alcohol use disorders identification Test (AUDIT)	Quantitative measure of alcohol consumption and problems	141,932	European	[81]
Maximum regular alcohol use	Quantitative measure of maximum reported habitual alcohol consumption	143,965	European and African	[82]

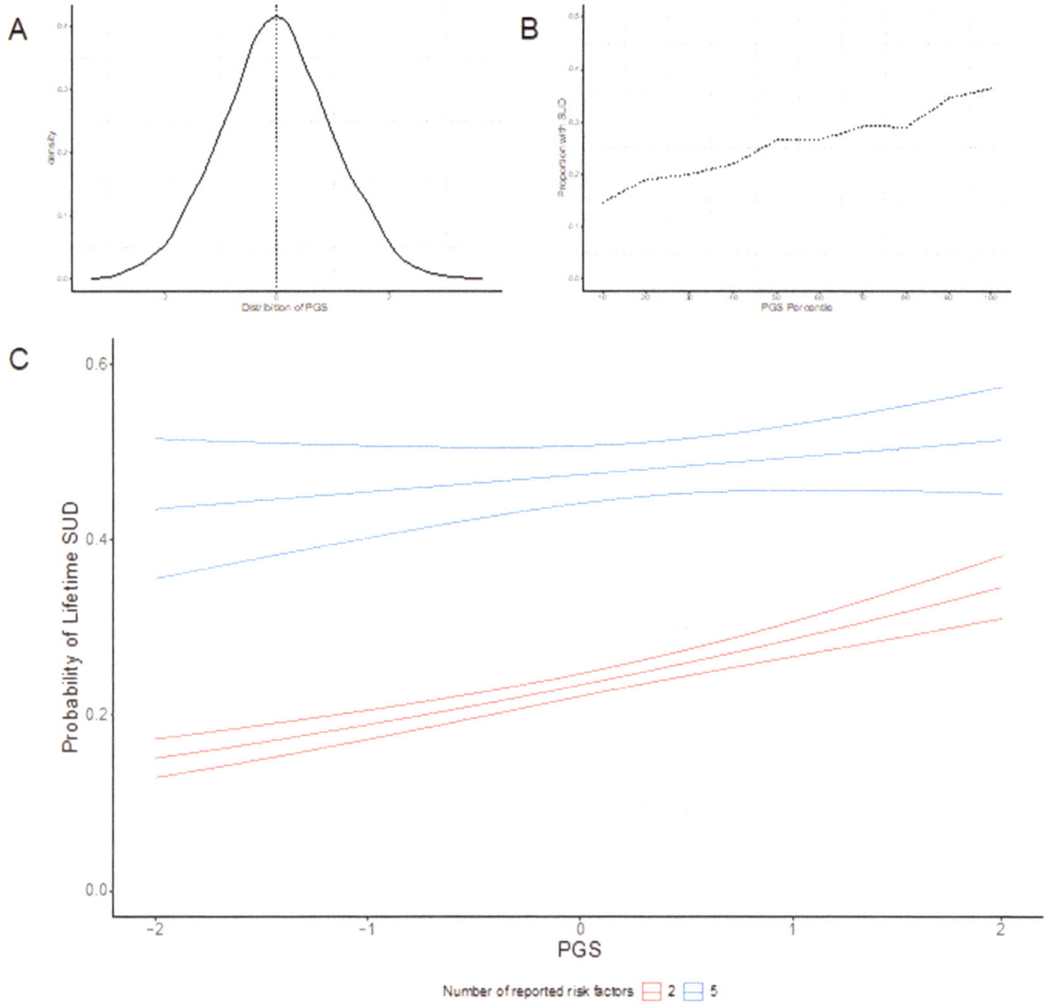

Fig. 3 Analysis of polygenic risk and lifetime SUD in Add Health

SE = 0.034, $p = 2.05 \times 10^{-26}$, *pseudo*-R^2 = 3.41%) in models adjusted for age, sex, and the first ten ancestral PCs. The decile plot (*see* Fig. 3, Panel b) shows a relatively monotonic increase in the prevalence of lifetime SUD across the range of PGS deciles. We further explore how the association changes across various contextual factors (e.g., GxE), finding evidence of an interaction between an index of validated risk factors for SUD [63] and the PGS (OR = 0.934, SE = 0.022, $p = 1.87 \times 10^{-3}$). Figure 3 (Panel c) helps to visualize how the PGS association is stronger for those who report fewer risk factors compared to those who report many. When performing analyses pertaining to GxE, it is also important to consider whether results are robust to covariate-by-PGS and covariate-by-environment interactions [64].

2.3 Limitations of Current Approaches

While new methods like polygenic scores are being successfully used in research, they are not without limitations. Incorporating polygenic scores into clinical practice has become a recent topic of interest [65–67], but the current evidence for alcohol and other substance use disorders suggests these tools are not ready for clinical use in any meaningful way [63, 68, 69]. Predictive power of PGS is driven almost entirely by the sample size of discovery GWAS [70], so even larger sample size will be necessary to discover the majority of SNPs associated with AUD. Importantly, even if polygenic scores had sufficient power to identify those at risk of AUD, most of the research into the genetic architecture of psychiatric traits is almost exclusively in persons of European ancestries [37]. Polygenic scores derived from one population do not transport well across populations due to systematic differences in allele frequencies and patterns of LD [71, 72]. Methodological tools that can use information from multiple ancestries to improve cross-population prediction of PGS, such as PRS-CSx [73] and PolyPred [74], are available. However, these tools alone cannot overcome the lack of statistical power in currently available samples of non-European ancestries. If the goal of precision medicine is to incorporate genetic information into clinical practice, currently available information will make existing health disparities worse, rather than closing existing gaps [75]. There is considerable work that needs to be completed before this gap in GWAS discovery sample size is closed. Third, PGS tell us little about the underlying biology of a disorder. Methods have been developed that take biological information into consideration when constructing PGS [76], but the focus remains on using that information to increase predictive power. Lastly, none of the limitations above address the possible ethical dilemmas related to the clinical and/or public use of genetic information that is highly probabilistic, compared to well-characterized, monogenic variants of strong effect (e.g., Huntington's disease). The field has made immense progress over recent decades regarding the genetic epidemiology of AUD, but there is still much to be done.

3 Conclusions

Alcohol use disorders are due, at least in part, to genetic influences. In this chapter, we have laid out several methods for latent and measured genetic approaches that allow researchers to incorporate this information into their research design. Latent methods, including using family and twin data, can be useful for understanding the variance in the population due to additive genetic influences, while designs that infer risk from affected individuals with known biological relatedness help to better characterize individual risk. Measured genotypic designs, such as genome-wide association studies and polygenic score analyses help us to understand variants associated with AUD and provide aggregate measures of individual risk, respectively. While all these approaches are useful for research, there is still considerable work to be done before they are ready for any clinical application. Additionally, genetics only tell part of the story, so information on social, environmental, and other risk factors is also important.

References

1. Degenhardt L, Charlson F, Ferrari A et al (2018) The global burden of disease attributable to alcohol and drug use in 195 countries and territories, 1990–2016: a systematic analysis for the Global Burden of Disease Study 2016. Lancet Psychiatry 5:987–1012. https://doi.org/10.1016/S2215-0366(18)30337-7
2. Murray CJLL, Mokdad AH, Ballestros K et al (2018) The state of US health, 1990–2016: burden of diseases, injuries, and risk factors among US states. JAMA 319:1444–1472. https://doi.org/10.1001/jama.2018.0158
3. Vigo D, Thornicroft G, Atun R (2016) Estimating the true global burden of mental illness. Lancet Psychiatry 3:171–178. https://doi.org/10.1016/S2215-0366(15)00505-2
4. Griswold MG, Fullman N, Hawley C et al (2018) Alcohol use and burden for 195 countries and territories, 1990–2016: a systematic analysis for the Global Burden of Disease Study 2016. Lancet 392:1015–1035. https://doi.org/10.1016/S0140-6736(18)31310-2
5. Smothers BA, Yahr HT, Ruhl CE (2004) Detection of alcohol use disorders in general hospital admissions in the United States. Arch Intern Med 164:749–756. https://doi.org/10.1001/ARCHINTE.164.7.749
6. Mekonen T, Chan GCK, Connor J et al (2021) Treatment rates for alcohol use disorders: a systematic review and meta-analysis. Addiction 116:2617–2634. https://doi.org/10.1111/ADD.15357
7. Galea S, Nandi A, Vlahov D (2004) The social epidemiology of substance use. Epidemiol Rev 26:36–52. https://doi.org/10.1093/EPIREV/MXH007
8. Zhou H, Sealock JM, Sanchez-Roige S et al (2020) Genome-wide meta-analysis of problematic alcohol use in 435,563 individuals yields insights into biology and relationships with other traits. Nat Neurosci. https://doi.org/10.1038/s41593-020-0643-5
9. Karriker-Jaffe KJ (2013) Neighborhood socioeconomic status and substance use by U.S. adults. Drug Alcohol Depend 133:212–221. https://doi.org/10.1016/j.drugalcdep.2013.04.033
10. Verhulst B, Neale MC, Kendler KS (2015) The heritability of alcohol use disorders: a meta-analysis of twin and adoption studies. Psychol Med 45:1061–1072. https://doi.org/10.1017/S0033291714002165
11. Sanchez-Roige S, Palmer AA, Clarke TK (2020) Recent efforts to dissect the genetic basis of alcohol use and abuse. Biol Psychiatry 87(7):609–618. https://doi.org/10.1016/j.biopsych.2019.09.011
12. Walters RK, Polimanti R, Johnson EC et al (2018) Trans-ancestral GWAS of alcohol dependence reveals common genetic underpinnings with psychiatric disorders. Nat Neurosci 21:1656–1669. https://doi.org/10.1038/s41593-018-0275-1
13. Karlsson Linnér R, Mallard TT, Barr PB et al (2021) Multivariate analysis of 1.5 million people identifies genetic associations with traits related to self-regulation and addiction. Nat Neurosci:1–10. https://doi.org/10.1038/s41593-021-00908-3
14. Demontis D, Walters RK, Martin J et al (2019) Discovery of the first genome-wide significant risk loci for attention deficit/hyperactivity disorder. Nat Genet 51:63–75. https://doi.org/10.1038/s41588-018-0269-7
15. Zhou H, Rentsch CT, Cheng Z et al (2020) Association of OPRM1 functional coding variant with opioid use disorder: a genome-wide association study. JAMA Psychiatry. https://doi.org/10.1001/jamapsychiatry.2020.1206
16. Johnson EC, Demontis D, Thorgeirsson TE et al (2020) A large-scale genome-wide association study meta-analysis of cannabis use disorder. Lancet Psychiatry. https://doi.org/10.1016/S2215-0366(20)30339-4

17. Levey DF, Stein MB, Wendt FR et al (2021) Bi-ancestral depression GWAS in the million veteran program and meta-analysis in >1.2 million individuals highlight new therapeutic directions. Nat Neurosci. https://doi.org/10.1038/s41593-021-00860-2
18. Levey DF, Gelernter J, Polimanti R et al (2020) Reproducible genetic risk loci for anxiety: results from ~200,000 participants in the million veteran program. Am J Psychiatry. https://doi.org/10.1176/appi.ajp.2019.19030256
19. Mullins N, Forstner AJ, O'Connell KS et al (2021) Genome-wide association study of more than 40,000 bipolar disorder cases provides new insights into the underlying biology. Nat Genet. https://doi.org/10.1038/s41588-021-00857-4
20. Trubetskoy V, Pardiñas AF, Qi T et al (2022) Mapping genomic loci implicates genes and synaptic biology in schizophrenia. Nature 2022:1–13. https://doi.org/10.1038/s41586-022-04434-5
21. Kendler KS, Gardner C, Dick DM (2011) Predicting alcohol consumption in adolescence from alcohol-specific and general externalizing genetic risk factors, key environmental exposures and their interaction. Psychol Med 41:1507–1516. https://doi.org/10.1017/S003329171000190X.Predicting
22. Meyers JL, Salvatore JE, Vuoksimaa E et al (2014) Genetic influences on alcohol use behaviors have diverging developmental trajectories: a prospective study among male and female twins. Alcohol Clin Exp Res 38:2869–2877. https://doi.org/10.1111/acer.12560
23. McLeod JD (1995) Social and psychological bases of homogamy for common psychiatric disorders. J Marriage Fam 57:201. https://doi.org/10.2307/353828
24. Barr PB, Kuo SI, Aliev F et al (2019) Polygenic risk for alcohol misuse is moderated by romantic partnerships. Addiction 114:1753–1762. https://doi.org/10.1111/add.14712
25. Stoltenberg SF, Mudd SA, Blow FC, Hill EM (1998) Evaluating measures of family history of alcoholism: density versus dichotomy. Addiction 93:1511–1520. https://doi.org/10.1046/J.1360-0443.1998.931015117.X
26. Pandey G, Seay MJ, Meyers JL et al (2020) Density and dichotomous family history measures of alcohol use disorder as predictors of behavioral and neural phenotypes: a comparative study across gender and race/ethnicity. Alcohol Clin Exp Res 44:697–710. https://doi.org/10.1111/ACER.14280
27. Bross IDJ (1958) How to use ridit analysis. Biometrics 14:18. https://doi.org/10.2307/2527727
28. Wichers M, Gillespie NA, Kendler KS (2013) Genetic and environmental predictors of latent trajectories of alcohol use from adolescence to adulthood: a male twin study. Alcohol Clin Exp Res 37:498–506. https://doi.org/10.1111/J.1530-0277.2012.01939.X
29. Kendler KS, Ohlsson H, Sundquist J, Sundquist K (2021) Family genetic risk scores and the genetic architecture of major affective and psychotic disorders in a Swedish national sample. JAMA Psychiatry 78:735–743. https://doi.org/10.1001/JAMAPSYCHIATRY.2021.0336
30. Kendler KS, Ohlsson H, Sundquist J, Sundquist K (2021) The impact of sex, age at onset, recurrence, mode of ascertainment and medical complications on the family genetic risk score profiles for alcohol use disorder. Psychol Med:1–9. https://doi.org/10.1017/S0033291721003317
31. Kendler KS, Ohlsson H, Sundquist J, Sundquist K (2021) The patterns of family genetic risk scores for eleven major psychiatric and substance use disorders in a Swedish national sample. Transl Psychiatry 11:1–8. https://doi.org/10.1038/s41398-021-01454-z
32. Duncan LE, Keller MC (2011) A critical review of the first 10 years of candidate gene-by-environment interaction research in psychiatry. Am J Psychiatr 168:1041–1049. https://doi.org/10.1176/appi.ajp.2011.11020191
33. Border R, Johnson EC, Evans LM et al (2019) No support for historical candidate gene or candidate gene-by-interaction hypotheses for major depression across multiple large samples. Am J Psychiatr 176:376–387. https://doi.org/10.1176/appi.ajp.2018.18070881
34. Liu M, Jiang Y, Wedow R et al (2019) Association studies of up to 1.2 million individuals yield new insights into the genetic etiology of tobacco and alcohol use. Nat Genet 51:237–244. https://doi.org/10.1038/s41588-018-0307-5
35. Kranzler HR, Zhou H, Kember RL et al (2019) Genome-wide association study of alcohol consumption and use disorder in 274,424 individuals from multiple populations. Nat Commun 10:1499. https://doi.org/10.1038/s41467-019-09480-8
36. Visscher PM, Wray NR, Zhang Q et al (2017) 10 years of GWAS discovery: biology, function, and translation. Am J Hum Genet 101:5–22. https://doi.org/10.1016/j.ajhg.2017.06.005

37. Mills MC, Rahal C (2019) A scientometric review of genome-wide association studies. Commun Biol 2:9. https://doi.org/10.1038/s42003-018-0261-x
38. Verlouw JAM, Clemens E, de Vries JH et al (2021) A comparison of genotyping arrays. Eur J Hum Genet 29:1611–1624. https://doi.org/10.1038/s41431-021-00917-7
39. Uffelmann E, Huang QQ, Munung NS et al (2021) Genome-wide association studies. Nat Rev Methods Primers 1:1–21. https://doi.org/10.1038/s43586-021-00056-9
40. Auton A, Abecasis GR, Altshuler DM et al (2015) A global reference for human genetic variation. Nature 526:68–74. https://doi.org/10.1038/nature15393
41. Harris KM (2013) The Add Health Study: Design and Accomplishments. https://doi.org/10.17615/C6TW87
42. Harris KM (2009) The National Longitudinal Study of Adolescent to Adult Health (Add Health), waves I & II, 1994–1996; wave III, 2001–2002; wave IV, 2007–2009 [machine-readable data file and documentation]. Carolina Population Center, University of North Carolina at Chapel Hill, Chapel Hill
43. Braudt DB, Mullan Harris K (2018) Polygenic Scores (PGSs) in the National Longitudinal Study of Adolescent to Adult Health (Add Health) – release 1. Carolina Population Center, Chapel Hill
44. Peterson RE, Kuchenbaecker K, Walters RK et al (2019) Genome-wide association studies in ancestrally diverse populations: opportunities, methods, pitfalls, and recommendations. Cell 179:589–603. https://doi.org/10.1016/j.cell.2019.08.051
45. Peterson RE, Edwards AC, Bacanu SA et al (2017) The utility of empirically assigning ancestry groups in cross-population genetic studies of addiction. Am J Addict 26:494–501. https://doi.org/10.1111/ajad.12586
46. Taliun D, Harris DN, Kessler MD et al (2021) Sequencing of 53,831 diverse genomes from the NHLBI TOPMed program. Nature 590(7845):290–299. https://doi.org/10.1038/s41586-021-03205-y
47. Watanabe K, Taskesen E, van Bochoven A, Posthuma D (2017) Functional mapping and annotation of genetic associations with FUMA. Nat Commun 8:1–11. https://doi.org/10.1038/s41467-017-01261-5
48. Sey NYA, Hu B, Mah W et al (2020) A computational tool (H-MAGMA) for improved prediction of brain-disorder risk genes by incorporating brain chromatin interaction profiles. Nature Neuroscience 23(4):583–593. https://doi.org/10.1038/s41593-020-0603-0
49. Gamazon ER, Wheeler HE, Shah KP et al (2015) A gene-based association method for mapping traits using reference transcriptome data. Nature Genetics 47(9):1091–1098. https://doi.org/10.1038/ng.3367
50. Barbeira AN, Dickinson SP, Bonazzola R et al (2018) Exploring the phenotypic consequences of tissue specific gene expression variation inferred from GWAS summary statistics. Nat Commun 9(1):1–20. https://doi.org/10.1038/s41467-018-03621-1
51. Bulik-Sullivan B, Finucane HK, Anttila V et al (2015) An atlas of genetic correlations across human diseases and traits. Nat Genet 47(11):1236–1241. https://doi.org/10.1038/ng.3406
52. Bulik-Sullivan BK, Loh P-RR, Finucane HK et al (2015) LD Score regression distinguishes confounding from polygenicity in genome-wide association studies. Nat Genet 47:291–295. https://doi.org/10.1038/ng.3211
53. Finucane HK, Bulik-Sullivan B, Gusev A et al (2015) Partitioning heritability by functional annotation using genome-wide association summary statistics. Nat Genet 47:1228. https://doi.org/10.1038/ng.3404. https://www.nature.com/articles/ng.3404#supplementary-information
54. Sanderson E, Glymour MM, Holmes MV et al (2022) Mendelian randomization. Nat Rev Methods Primers 2(1):1–21. https://doi.org/10.1038/s43586-021-00092-5
55. Davey Smith G, Ebrahim S, Smith GD, Ebrahim S (2004) Mendelian randomization: prospects, potentials, and limitations. Int J Epidemiol 33:30–42. https://doi.org/10.1093/ije/dyh132
56. Davey Smith G, Hemani G (2014) Mendelian randomization: genetic anchors for causal inference in epidemiological studies. Hum Mol Genet 23:R89–R98. https://doi.org/10.1093/hmg/ddu328
57. Lawlor DA, Harbord RM, Sterne JACC et al (2008) Mendelian randomization: using genes as instruments for making causal inferences in epidemiology. Stat Med 27:1133–1163. https://doi.org/10.1002/sim.3034
58. Vilhjálmsson BJ, Jian Yang HKF, Alexander Gusev SL et al (2015) Modeling linkage disequilibrium increases accuracy of polygenicrisk scores. Am J Hum Genet 97:576–592. https://doi.org/10.1016/j.ajhg.2015.09.001
59. Ge T, Chen C-Y, Ni Y et al (2019) Polygenic prediction via Bayesian regression and

continuous shrinkage priors. Nat Commun 10: 1776. https://doi.org/10.1038/s41467-019-09718-5

60. Ni G, Zeng J, Revez JA et al (2021) A comparison of ten polygenic score methods for psychiatric disorders applied across multiple cohorts. Biol Psychiatry 90:611–620. https://doi.org/10.1016/J.BIOPSYCH.2021.04.018

61. Sohail M, Maier RM, Ganna A et al (2019) Polygenic adaptation on height is overestimated due to uncorrected stratification in genome-wide association studies. elife 8. https://doi.org/10.7554/ELIFE.39702

62. Berg JJ, Harpak A, Sinnott-Armstrong N et al (2019) Reduced signal for polygenic adaptation of height in UK biobank. elife 8. https://doi.org/10.7554/ELIFE.39725

63. Barr PB, Driver MN, I-Chun Kuo S et al (2022) Clinical, environmental, and genetic risk factors for substance use disorders: characterizing combined effects across multiple cohorts. Mol Psychiatry 27:4633–4641. https://doi.org/10.1038/s41380-022-01801-6

64. Keller MC (2013) Gene × environment interaction studies have not properly controlled for potential confounders: the problem and the (simple) solution. Biol Psychiatry 75:18–24. https://doi.org/10.1016/j.biopsych.2013.09.006

65. Lewis CM, Vassos E (2020) Polygenic risk scores: from research tools to clinical instruments. Genome Med 12:1–11. https://doi.org/10.1186/S13073-020-00742-5/TABLES/2

66. Wray NR, Lin T, Austin J et al (2021) From basic science to clinical application of polygenic risk scores. JAMA Psychiatry 78(1): 101–109. https://doi.org/10.1001/jamapsychiatry.2020.3049

67. Hao L, Kraft P, Berriz GF et al (2022) Development of a clinical polygenic risk score assay and reporting workflow. Nat Med 28(5): 1006–1013. https://doi.org/10.1038/s41591-022-01767-6

68. Barr PB, Ksinan A, Su J et al (2020) Using polygenic scores for identifying individuals at increased risk of substance use disorders in clinical and population samples. Transl Psychiatry 10:196. https://doi.org/10.1038/s41398-020-00865-8

69. Yeung EW, Spychala KM, Miller AP et al (2022) Effects of genetic risk for alcohol dependence and onset of regular drinking on the progression to alcohol dependence: a polygenic risk score approach. Drug Alcohol Depend 230:109117. https://doi.org/10.1016/J.DRUGALCDEP.2021.109117

70. Dudbridge F (2013) Power and predictive accuracy of polygenic risk scores. PLoS Genet 9:e1003348. https://doi.org/10.1371/journal.pgen.1003348

71. Duncan L, Shen H, Gelaye B et al (2019) Analysis of polygenic risk score usage and performance in diverse human populations. Nat Commun. https://doi.org/10.1038/s41467-019-11112-0

72. Martin AR, Gignoux CR, Walters RK et al (2017) Human demographic history impacts genetic risk prediction across diverse populations. Am J Hum Genet 100:635–649. https://doi.org/10.1016/j.ajhg.2017.03.004

73. Ruan Y, Lin Y-F, Feng Y-CA et al (2022) Improving polygenic prediction in ancestrally diverse populations. Nat Genet. https://doi.org/10.1038/s41588-022-01054-7

74. Weissbrod O, Kanai M, Shi H et al (2022) Leveraging fine-mapping and multipopulation training data to improve cross-population polygenic risk scores. Nat Genet 54(4):450–458. https://doi.org/10.1038/s41588-022-01036-9

75. Martin AR, Kanai M, Kamatani Y et al (2019) Clinical use of current polygenic risk scores may exacerbate health disparities. Nat Genet 51:584–591. https://doi.org/10.1038/s41588-019-0379-x

76. Márquez-Luna C, Gazal S, Loh PR et al (2021) Incorporating functional priors improves polygenic prediction accuracy in UK biobank and 23andMe data sets. Nature Communications 12(1):1–11. https://doi.org/10.1038/s41467-021-25171-9

77. International Schizophrenia Consortium, Purcell SM, Wray NR et al (2009) Common polygenic variation contributes to risk of schizophrenia and bipolar disorder. Nature 460:748–752. https://doi.org/10.1038/nature08185

78. Privé F, Arbel J, Vilhjálmsson BJ (2020) LDpred2: better, faster, stronger. Bioinformatics 36:1–14. https://doi.org/10.1093/bioinformatics/btaa1029

79. Lloyd-Jones LR, Zeng J, Sidorenko J et al (2019) Improved polygenic prediction by Bayesian multiple regression on summary statistics. Nat Commun 10(1):1–11. https://doi.org/10.1038/s41467-019-12653-0

80. Mak TSH, Porsch RM, Choi SW et al (2017) Polygenic scores via penalized regression on

summary statistics. Genet Epidemiol 41:469–480. https://doi.org/10.1002/GEPI.22050

81. Sanchez-Roige S, Palmer AA, Fontanillas P et al (2019) Genome-wide association study meta-analysis of the alcohol use disorders identification test (AUDIT) in two population-based cohorts. Am J Psychiatr 176:107–118. https://doi.org/10.1176/appi.ajp.2018.18040369

82. Gelernter J, Sun N, Polimanti R et al (2019) Genome-wide association study of maximum habitual alcohol intake in >140,000 U.S. European and African American veterans yields novel risk loci. Biol Psychiatry 86:365–376. https://doi.org/10.1016/j.biopsych.2019.03.984

Chapter 2

Pavlovian Spontaneous Recovery as a Measure of Alcohol-Seeking Behavior

Sheketha R. Hauser, Richard L. Bell, and Zachary A. Rodd

Abstract

Alcohol (EtOH) addiction can be characterized as a chronic relapsing disorder. EtOH craving is a critical precipitating factor to relapse. Environmental and conditioning cues can elicit drug-seeking 'craving' in both humans and rodents. The operant Pavlovian Spontaneous Recovery (PSR) model of EtOH-seeking is a unique and suitable model for studying 'craving-like' behavior as well as examining underlying neural systems and mechanisms that may be involved in 'craving-like' and, by extension, relapse behavior in rodents. Our objective is to provide the reader with details and methods on the benefits of using the PSR model to examine and better understand EtOH 'craving-like' behavior.

Key words Pavlovian Spontaneous recovery, Alcohol-seeking, Craving, Alcohol

1 Introduction

Drug addiction can be characterized as a chronic relapsing disorder. The relapse rate of alcoholism is particularly high [1]. For example, alcohol (EtOH) use disorders (AUDs) being treated in a cognitive behavioral treatment setting have a low success rate at 3- to 6-month follow-ups [1]. Within the first 6 months of Alcoholics Anonymous (AA) attendance, most individuals starting the program quit, with the relapse rate being around 85% [1]. The chronic relapsing nature of AUDs is one of the major challenges in addiction treatment.

In general, the probability of relapse to drug intake increases over time, which interferes with maintaining sobriety [2]. Drug craving is a critical precipitating factor to relapse [3]. As the passage of time increases during abstinence, drug-craving levels increase over time, which can concomitantly increase the likelihood of relapse [3]. Although, the trajectory follows an inverted U-shaped curve. For instance, Li et al. [4] reported that EtOH-craving was significantly higher following 60 days of abstinence

Melissa A. Cyders (ed.), *Translational Research Methods for Alcohol Use Disorders*, Neuromethods, vol. 201,
https://doi.org/10.1007/978-1-0716-3267-3_2,

compared to 7, 14, or 30 days of abstinence. Furthermore, cue-induced EtOH craving was also highest following 60 days of abstinence [4]. Rehabilitation services are typically short (e.g., ~30 days); therefore, completion of these programs can occur before peak EtOH craving occurs, such that the severity of craving at the end of treatment can predict relapse at 90-days [5].

1.1 Spontaneous Recovery

In 1927, Pavlov reported on the phenomenon he termed spontaneous recovery [6]. His work demonstrated that following several extinction trials, test subjects would salivate in response to the ringing bell if there was a sufficient passage of time from the last extinction session [7, 8]. This phenomenon is called spontaneous recovery, defined as the reinstatement of responding (goal seeking) or a conditional response in the absence of the previously trained reward following a period of rest after extinction [7]. The spontaneous recovery phenomenon has also been observed in drug addiction. For example, DiCiano and Everitt's [9] findings indicated that spontaneous recovery predicts abstinent drug-dependent patients' risk of relapse, which increases with time [9, 10]. Spontaneous recovery of drug-seeking can also be triggered by environmental context and drug cues within this context. For example, drug cues can induce cravings for cocaine in addicts following a 12-month period of abstinence [10, 11].

1.2 Pavlovian Spontaneous Recovery

Operant conditioning is a method of learning that occurs through the association of a behavioral response and the consequence of that response. It has proven to be a very valuable scientific tool used to examine the rewarding effects of drug-associated behaviors such as self-administration (drug-intake), progressive ratio breakpoints (assessing maximum amount of work or motivation to acquire the drug), extinction (assessing discontinuation of the response when the drug is absent), drug-seeking (assessing craving-like behavior), and relapse (assessing reacquisition of drug intake) [12–17].

Various drug-seeking models have proven to be valuable tools for elucidating the neural systems mediating and mechanisms for this mediation underlying drug-seeking behavior. For example, drug priming (priming injection of the previously self-administered drug; [18–20]), exposure to a physical stressor (e.g., footshock) or creating a physical state of extreme distress (yohimbine) (stress-induced drug-seeking; [21, 22]), incubation of drug craving (a progressive increase of cue-induced drug craving during forced abstinence from drug self-administration; [23, 24]), and the renewal model (contextual drug-seeking; self-administer a drug in a particular environment [A], the behaviors are extinguished in a different context [B], and responding is returned when the animal is returned to the original context [A]; [25–29]) and cue-induced drug-seeking (presentation of the previously paired cue with the availability of the associated drug; [30, 31]).

Since spontaneous recovery is defined in clinical treatment literature as the cessation of alcohol consumption in human alcoholics without an overt treatment intervention, we have modified the term to "Pavlovian Spontaneous Recovery" (PSR) as a measure of EtOH-seeking behavior [7]. The PSR is a unique phenomenon in that it is time-dependent, and the behavior appears to be dependent on the re-exposure of the organism to all the cues in the behavioral environment previously associated with the reinforcer [7]. Thus, the PSR model is a contextual/environmental stimulus model of EtOH-seeking behavior. Furthermore, the expression of a PSR is directly correlated to reward salience [32, 33]. Also, the PSR measures EtOH-seeking after observable physical signs of withdrawal have dissipated. Thus, the PSR model may be a measure of 'craving-like' behavior, in addition to a strictly EtOH-seeking behavioral model of operant responding in the absence of a reinforcer [10].

The PSR has been demonstrated in alcohol-preferring (P) rats [34–43] and non-selective rats as well [44]. To validate the PSR model, the efficacy of an FDA-approved anti-craving agent, naltrexone, was tested in this model, and naltrexone was found to decrease the expression of EtOH PSR, thus confirming the utility of PSR as a model of EtOH-seeking behavior and an increased probability of relapse [7, 38]. Thus, the persistent expression of PSR in the absence of reward is thought to resemble the compulsive nature of drug abuse seen in humans [7, 45, 46]. Moreover, the magnitude of a PSR response increases with the passage of time since the last extinction session (more than 1 week; see [7, 8, 47]), which suggests that the expression of an EtOH PSR-response is mediated by neuroadaptations associated with EtOH-seeking, and putatively craving, and not the result of drug priming [8]. It is postulated that PSR reflects a shift from recent extinction (second-order) learning, which builds cumulatively across extinction sessions, to the initial drug-reinforced response, reflecting an intrinsic motivation to obtain the previously administered reward. This suggests that PSR is a model suitable for studying 'craving-like' behavior [7, 8, 37, 38, 44], in and of itself, as well as examining underlying neural systems and mechanisms that are involved in 'craving-like' and other relapse-associated behaviors (e.g., breakpoint; [37, 40–42, 48–50]).

2 Materials

2.1 Animal Preparation

For our experiments, we used adult EtOH naïve alcohol-preferring (P) rats (male or female) at ≥ postnatal day (PND) 90, weighing 250–400 g (female and male, respectively) at the start of the experiment. P rats are selectively bred in-house at Indiana University School of Medicine, Indianapolis, and will readily drink pharmacologically relevant levels of EtOH as well as readily acquire operant EtOH self-administration [51]. The rats were double housed and maintained on a 12-h reversed light-dark cycle (lights off at 0900 h) because rats are nocturnal and are, thus, more active during the dark phase. Therefore, we operantly train and test rats during their active phase to measure more natural behavioral responses. Food and water are available throughout the experiment in the home cage. Rats were handled 1 week prior to the operant sessions for 5–10 min/rat to minimize any handling stress, and their tails are numbered weekly to identify the rats.

Unlike naïve P rats, naïve non-selected rats (e.g., Sprague Dawley, Wistar, Long-Evans Hooded) do not readily acquire EtOH-self administration under operant conditions (i.e., require some form of shaping or fading procedure) due to the aversive taste of alcohol [52]. The animals need to have stable, meaningful levels of EtOH-self-administration prior to measuring EtOH-seeking behaviors [52]. Therefore, these outbred animals are trained to drink EtOH prior to operant self-administration via water deprivation, intermittent-access 2-bottle [44], limited access [53], sucrose-fading [30, 54, 55], or other EtOH drinking procedures (e.g., [56]) In addition, Augier et al. [57] demonstrated that Wistar rats will self-administer 20% EtOH without sucrose or saccharin fading or any water deprivation, thus increasing EtOH to a 20% concentration instead of using 10% EtOH in EtOH operant conditioning was shown to improve Wistar rats' ability to acquire of EtOH self-administration [52]. Thus, there are EtOH-drinking and operant training techniques that will increase outbred rats' acquisition and maintenance to self-administer EtOH under operant conditions.

2.2 Chemical Agents and Vehicle

Ethyl alcohol (190 proof; McCormick Distilling Co., Weston, MO) was diluted to 15% with distilled water or reverse osmosis (RO) water for operant oral self-administration. 200 proof EtOH is avoided since the last 5% of impurities are chemically "burnt off" post-distillation.

2.3 Operant Apparatus

EtOH self-administration experiments are conducted in standard two-lever operant chambers (e.g., Coulbourn Instruments, Allentown, PA) contained within ventilated, sound-attenuated enclosures [34, 35, 40–43]. Other operant systems include Med Associates (St. Albans, VT) (*see* [44]), San Diego Instruments (SDI: San Diego, CA 92121), and others. Two operant levers were located on the same wall and were placed 15 cm above a grid floor and 13 cm apart. Directly beneath each lever was a trough through which a dipper cup (0.1 mL) rose to deliver response contingent fluid. Upon a reinforced response, a small cue light was illuminated in the drinking trough during the 4-s dipper cup access. The assignment of water and EtOH lever with respect to the left or right position was counterbalanced among subjects. However, the EtOH and water levers remained the same for each rat. Throughout the experimental sessions, the operant chambers were illuminated by a house light. A microcomputer controlled the operant chambers and recorded data. All sessions were 60 min in duration and were conducted daily (Monday–Sunday) during the reverse light cycle.

2.4 Operant Training

The PSR model of EtOH-seeking follows a procedure in which self-administration occurs in context A, extinction in context A, a home cage period with no experimenter manipulation, and drug-seeking upon returning to context A [34, 35]. For this review, we will go through each step. For P rats, ethyl alcohol (190 proof; McCormick Distilling Co., Weston, MO) is diluted to 15% with distilled water or reverse osmosis (RO) water for operant oral self-administration. The operant troughs, which the animals cannot reach rather the dipper is raised upon a valid response, are filled with either 15% EtOH or water (e.g., distilled or reverse osmosis [RO]), and the operant computer program is initialized before the animals are removed from the vivarium. The rats are then brought to the operant room, and naïve rats are placed into the operant chamber without prior training.

(1) Operant sessions are 60 min in duration and occur daily for 10 weeks (i.e., Monday–Sunday; [34, 35, 40–43, 47, 48]: *see* Fig. 1 Created with BioRender.com). During the initial 4 weeks of daily operant access, both solutions (water; distilled or RO and 15% EtOH) are reinforced on a Fixed ratio-1 (FR-1) schedule, where one bar press on the associated lever results in one dipper presentation. At the end of the 4 weeks, the response requirement for EtOH dipper access is increased to an FR-3 schedule for 3 weeks, and then to FR-5 schedule for 3 weeks. Increasing the FR schedule for EtOH demonstrates that animal is motivated to attain EtOH because they will work for it. Water remains on an FR-1 schedule throughout the operant experiment. Previous studies have shown the average EtOH intake of P rats is approximately 1.2 g/kg/session. This level of intake results in blood EtOH levels of 60–80 mg% [51].

(2) After the rats have established stable levels of responding, across days, on the FR5 schedule for EtOH, they will undergo 7 days of extinction (60 min/day) when neither water nor EtOH is available ([40–43, 47, 48]; Fig. 1). Water is not available during the extinction procedure since water is a primary reinforcer and has been shown to influence responding during extinction testing, i.e., superstitious behavior [32]. Except for no fluid being present, the delivery system operates exactly as the preceding EtOH self-administration sessions. Extinction is present when responding on the previously paired EtOH lever is equivalent to that observed on the water lever, which usually requires seven to ten sessions [34, 35].

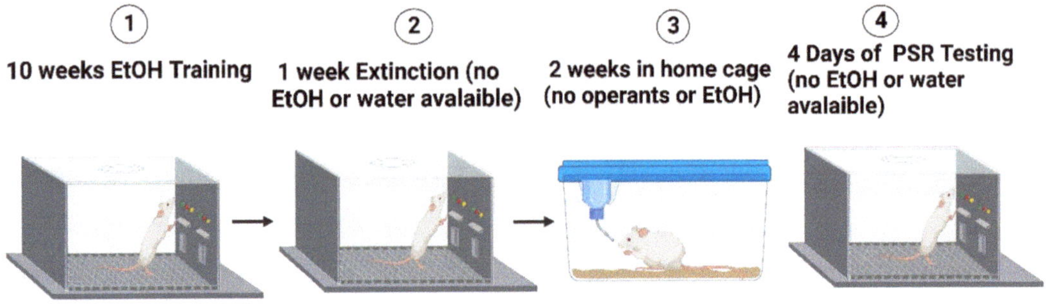

Fig. 1 Experimental timelines for PSR model of EtOH-seeking

2.5 Home Cage/Abstinence Phase

(3) After the last extinction session, the rats are maintained in their respective home cage for 2 weeks. In the home cage, the animals are not exposed to the operant chamber nor EtOH access nor any other experimenter manipulation (*see* Fig. 1). Two weeks of home cage rest has been shown to produce a robust PSR [34, 35, 48]. As previously mentioned, a temporal analysis, where length of home cage rest was manipulated, of EtOH-seeking in P rats demonstrated that 1 week of abstinence/home cage was insufficient for the expression of this EtOH-seeking behavior (PSR) in P rats. In contrast, following 2, 4, 6, or 8 weeks of abstinence, EtOH-seeking behavior was observed with the highest level of seeking displayed at 6 weeks of abstinence (*see* Fig. 2). Thus, providing further evidence that as time increases since the last extinction session, EtOH-seeking or "craving-like" behavior increases over time. These findings parallel clinical observations in abstinent AUD patients [3, 4, 7].

2.6 Pavlovian Spontaneous Recovery (PSR) Test

(4) After 2 weeks in the home cage, rats will undergo the PSR test (*see* Fig. 1). The PSR sessions are identical to the operant extinction protocol sessions. The FR5–FR1 schedule, lever contingencies, and dipper functioning are maintained, but EtOH and water are, again, absent. Rats are given four consecutive PSR test sessions because previous studies have shown that PSR-responding of adult P rats starts to return to baseline by the second session [34, 35, 48]. In addition, exposure to EtOH odor cues or EtOH priming doses or taste cues, or odor cues [34, 35, 39, 43], or particular pharmaceuticals [37, 41] may prolong PSR responding.

3 Testing Pharmacological Agents in the PSR Model

We have tested the systemic effects of different pharmacological agents using the PSR procedure to examine the mechanisms that may underlie EtOH-seeking behavior. Our findings demonstrated that metabotropic glutamate (mGlu) receptor 2/3 (mGlu2/3) receptors

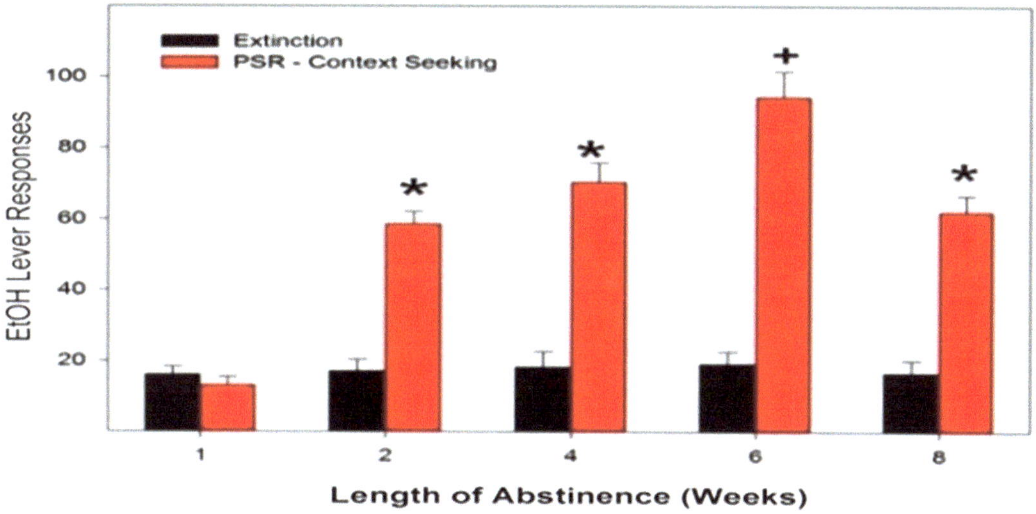

Fig. 2 PSR expression increases with a passage of time. (Modified from Hauser et al. [47]) Depicts the mean (± SEM) responses on the lever previously associated with the delivery of EtOH during PSR testing as a function of home cage duration. * Indicates that rats in the 2-, 4-, and 8-week groups responded more than their extinction baseline (Ext Base). + indicates that 6-week group's responding was elevated compared to extinction baseline and was significantly different than all other groups

(mGlu2/3agonist LY404039;), cannabinoid (CB) receptors 1 (CB1 antagonist SR141716A; [36]), non-selective opioid antagonists (naltrexone and LY25558; [38]), and selective kappa opioid receptors (kappa opioid antagonist JDTic; [49]) reduce EtOH-seeking behavior in the PSR model. However, orexin-1 receptors did not alter EtOH-seeking in the PSR model (orexin-1 receptor antagonist SB-334867; [37]). The PSR model can also be used to test the effects of other drugs-of-abuse on EtOH-seeking behavior. For example, systemic administration of cocaine [41] or nicotine [58] can enhance and/or reduce EtOH-seeking behavior in the PSR model.

The **(1)** operant training (10 weeks), **(2)** extinction training (1 week), and **(5)** PSR test session are the same as described above (*see* Fig. 3, Created with BioRender.com). The animals will be randomly assigned to a treatment group based on their previous self-administration pattern, such that each dose group will have equivalent response levels. The animals are to be handled during the last 4 days of the second week of home cage rest to acclimate the animals to intraperitoneal or subcutaneous injection. Handling of the animals lasts for 5–10 min, during which the animal is picked up and placed in the position in which the injection will be given; and, using a needleless syringe perform the injection as if it was real. Our experience is that this will minimize handling stress.

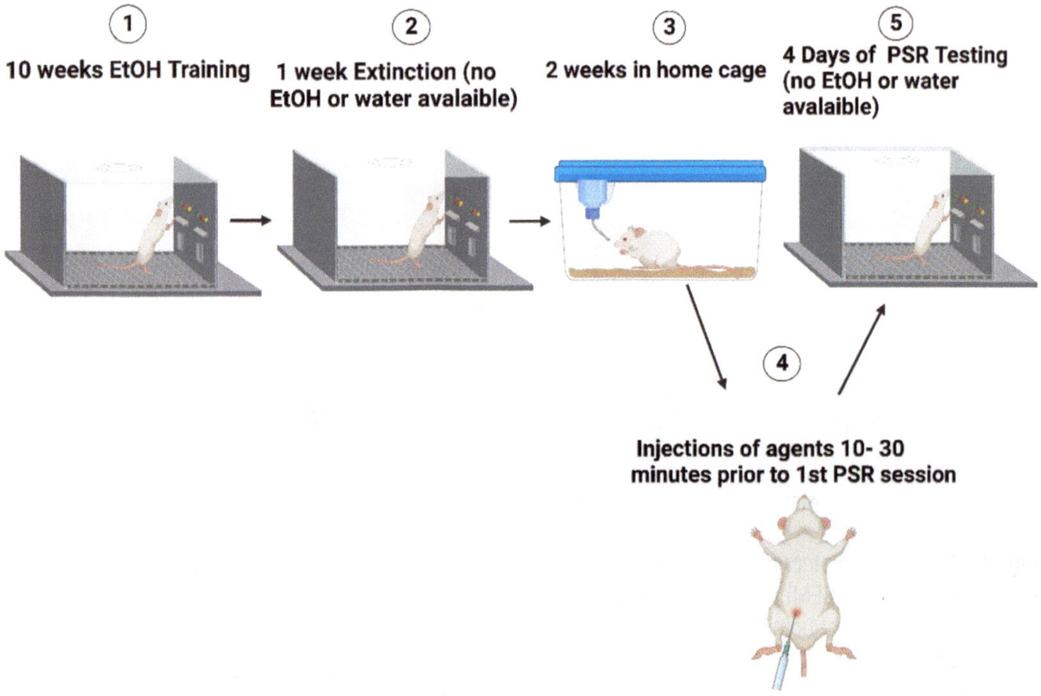

Fig. 3 Experimental timelines for systemic injections pharmacological agents during the PSR test

On PSR test day (single administration) or days (repeated administration), the rats are administered the vehicle or the drug. After about 10–30 min (this time will vary depending on the half-life of the drug or experimenters' protocol) drug administration, the animals are placed in the operant box for PSR testing.

4 Testing Pharmacological Agents in Discrete Brain Areas in the PSR Model

As previously mentioned, the PSR model can be used to examine underlying neural substrates that may be involved in EtOH-seeking behavior. The **(1)** operant training (10 weeks), **(2)** extinction training (1 week), and **(5)** PSR test session are the same as described above (*see* Fig. 4, Created with BioRender.com). During the eighth and ninth day of home cage rest, the animals are bilaterally implanted with a 22-gauge guide cannula (e.g., P1 Technologies, Roanoke, VA). Coordinates are chosen based on the Paxinos and Watson [59] stereotaxic rat atlas. For example, coordinates for placements to target the nucleus accumbens shell are +1.7 mm AP, +2.4 mm lateral to the midline, and −7.5 mm ventral from the surface of the skull at a 10° angle to the vertical; [59]. Institutional Animal Care and Use Committees (IACUCs) generally require record keeping of surgery and post-surgery procedures for each animal.

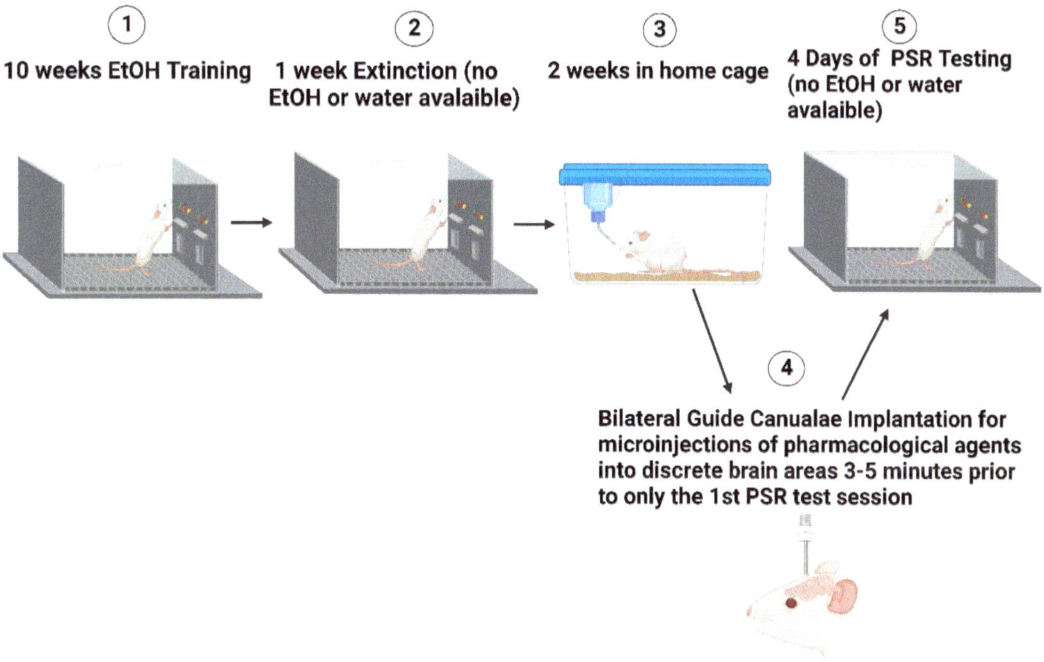

Fig. 4 Experimental timeline for microinjection of pharmacological agents into discrete brains areas prior to PSR testing

4.1 Surgery Preparation

The surgery area is disinfected before use. Hands are washed prior to donning surgical gloves. The surgeon will wear sterile surgical gloves, cap, mask, and clean lab coat or gown. The rats are anesthetized with an induction (e.g., using the nose-cone connected to a precision vaporizer) of 5% isoflurane before being transferred to a stereotaxic table. The animals will be maintained on a 2% isoflurane: air concentration. Hair over the incision site is shaved including a wide margin around the incision area. Skin is disinfected using several alternating rounds of surgical disinfectant scrub/solution (betadine, etc.) with 70% isopropyl alcohol rinses. An ophthalmic ointment is applied to each eye of the rat. Care is taken to avoid over-wetting fur outside of the surgical area. Instruments, sutures, wound clips, and implanted devices are sterilized prior to surgery, and a sterile field is maintained during surgery.

4.2 Microinjection Stereotaxic Surgery

(4) For stereotaxic surgery, after the rat is anesthetized, it is quickly transferred to the stereotaxic apparatus with the anesthesia level set at ~3% isoflurane. The rats are placed in a nose-cone set up for maintaining isoflurane anesthesia. After the rat has reached the appropriate anesthetic level (no tail pinch, footpad, or eye-blink reflexes), it is positioned in the stereotaxic unit with the ear bars and incisor bar (the latter is set at −3.3 mm). The reflexes are rechecked, and the level of anesthetic is lowered to 2% isoflurane.

Throughout surgery, rats are continuously observed for signs of respiratory distress, physical discomfort, and/or pain. Each incision site is thoroughly cleaned with surgical scrub. For stereotaxic surgeries, a ~2 to 3 cm incision is made to expose the skull over the target site(s). For bilateral injection experiments, two holes are bilaterally drilled through the skull to accommodate the bilateral implantation of 22-gauge guide cannula (e.g., P1 Technologies, Roanoke, VA). Two or three additional holes are drilled to accept cranial screws to ensure retention of the headcap. A long-lasting local analgesic (0.5% Bupivacaine) is infiltrated into all incision sites. Rats are also administered a subcutaneous injection of Carprofen (10 mg/kg).

Following the stereotaxic positioning of the guide cannula, it is permanently anchored to the cranial screws with cranioplastic acrylic cement. A 28-gauge stylet (e.g., P1 Technologies, Roanoke, VA) is placed into the guide cannulae and extended 0.5 mm beyond the tip of the guide to prevent clogging or any debris from getting into the cannula. After surgery, rats will be individually housed and will receive Carprofen (10 mg/kg/day) for 2 days post-surgery. The animals are allowed to recover for 7 days in their home cage. Animals are handled for at least 5 min daily beginning on the fourth recovery day and are habituated for two consecutive days to the mock injection procedures before microinjections prior to the first PSR test session.

4.3 Microinjection Procedure

On the first PSR test day, the drugs are dissolved into artificial cerebrospinal fluid (aCSF: consisting of 120.0 mM NaCl, 4.8 mM KCl, 1.2 mM KH_2PO_4, 1.2 mM Mg SO_4, 25.0 mM $NaHCO_3$, 2.5 mM $CaCl_2$, and 10.0 mM D-glucose). If the drug is not water-soluble and requires a different vehicle, then use the required vehicle as control and dissolve the drug in its respective vehicle. The drug and vehicle are microinjected with a Hamilton 25-μL syringe driven by a microinfusion syringe pump (e.g., Harvard Apparatus, MA, USA). After the drug has been freshly prepared and the microinjection system is set up, the animals will be brought from the home vivarium to the procedure room. 28-gauge injectors are inserted bilaterally to a depth of 1 mm beyond the end of the guide cannula and connected via tubing to a Hamilton 25-μL syringe driven by the microinfusion syringe pump (e.g., Harvard Apparatus, MA, USA) [40, 42]. A total drug-solution volume of 0.5 μL is administered over a 30 sec period per side; the injector tip is left in place for an additional 30 sec per side (note: the microinjection time, prior to the test session, may change depending on the drug administered). After microinjection of the drug, the animals are immediately placed into the operant chambers (*see* Fig. 4 Created with BioRender.com).

5 Method for Cue-Induced Expression of PSR

Pavlovian spontaneous recovery (PSR) is a model of context-induced seeking, but we can also use cues to enhance the expression of PSR. The addition of a non-specific conditioned cue (presence of EtOH odor) [35], EtOH-associated taste cues [39], white-noise conditioned stimulus [44], and other conditioned odor cues [43] can augment EtOH-seeking. Humans are exposed to conditioned cues in drug-paired and non-drug paired environments that trigger the craving for the previously used drug (e.g., pictures of beer steins, advertisements from EtOH distributors, and the odor of EtOH), but preclinical research has focused primarily on observations in the drug-paired environment (i.e., without the use of overt cues). We have utilized cues in two different ways (a) during the PSR test, the odor conditioned cues are presented in the operant chambers (*see* Fig. 5; Created with BioRender.com), or (b), when determining the effects of conditioned cues alone (CS^+ vs. CS^- vs. CS^0) we designed the experiment to divorce the conditioned cues from the drug-paired context and performance of drug self-administration associated behaviors (operant responding) by exposing the animals to the conditioned odors in a neutral environment for 20 min prior to PSR testing (*see* Fig. 6; Created with BioRender.com).

5.1 Cue Exposure

The **(1)** operant training (10 weeks), **(2)** extinction training (1 week), and **(4)** home cage rest are the same as described above (*see* Fig. 5). However, we will add the odor conditioning cues. We used three non-alcoholic odors for the three conditioned stimuli (e.g., peppermint, banana, and almond). **(1)** The excitatory condition stimulus (CS^+) will be paired with the 10 weeks of EtOH self-administration in the operant conditioning chambers (*see* Fig. 5; [43]). **(2)** The inhibitory condition stimulus CS^- odor is paired with the seven extinction sessions to increase cue-salience (extinction cue; *see* Fig. 5). Providing extinction reminders in clinical treatment to remind high-risk users of their therapeutic experiences inside the clinic may help decrease the probability of substance use in the outside world [60]. Therefore, we examine the effects of extinction cues in the PSR model of EtOH-seeking behavior. Clinically and pre-clinically, extinction cues have reduced EtOH and other drug-seeking behaviors [39, 43, 60–67].

During conditioning, conditioned odor cues for CS+ and CS− are emitted from a treated cotton ball (2–3 drops of the odor) placed outside the operant chamber, but within the sound attenuating chamber housing the operant chamber (*see* Fig. 5). **(3)** Neutral CS (CS^0) exposure will occur in a unique, non-drug-paired environment for 60 min, which occurs daily 60 min before the operant conditioning session, during the last 2 weeks of EtOH

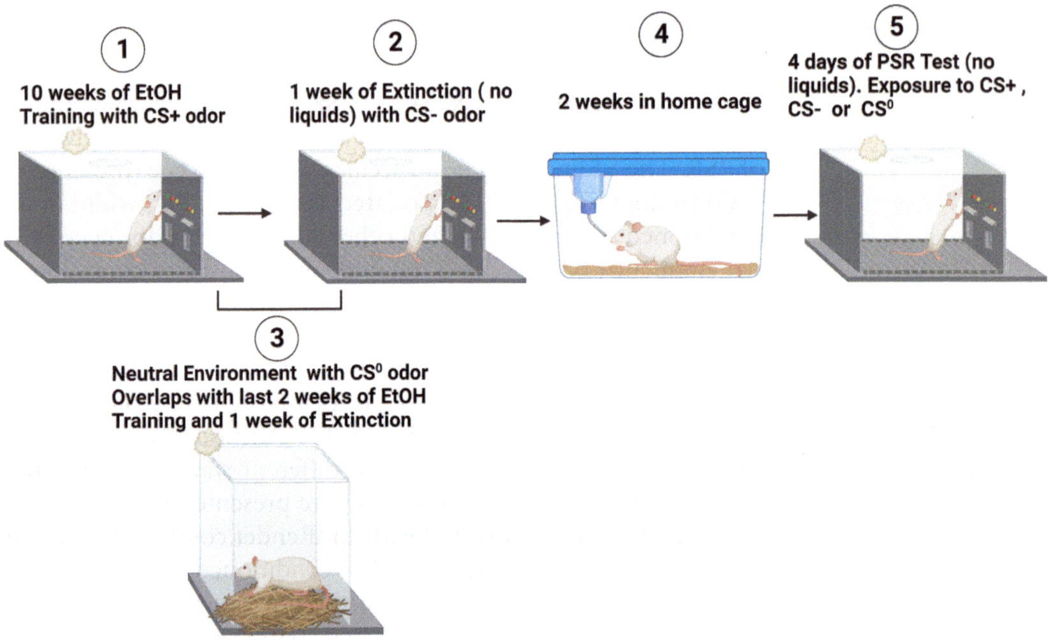

Fig. 5 Timeline for testing odors cues inside of the operant chambers

self-administration and 1 week of extinction (*see* Fig. 5). This is done, so the animals do not associate the neutral odor condition stimulus with either EtOH-self administration or extinction per se. The CS^0 is emitted from a treated cotton ball (2–3 drops of the odor) placed on top of a Plexiglass cage.

5.2 PSR Testing with Conditioned Odor Cues

After extinction training, all rats are maintained in the home cages for 2 weeks without operant or EtOH exposure (*see* Fig. 5; Created with BioRender.com). For the PSR test session, the rats are randomly assigned to experience a conditioned stimulus (CS^+, CS^0,or CS^-). **(5)** The assigned conditioned stimulus (CS^+, CS^0, or CS^-) odors are emitted from a treated cotton ball (2–3 drops of the odor) placed outside the operant chamber, but within the sound attenuated chamber housing the operant chamber. The PSR sessions are 60 min in duration (*see* Fig. 5).

5.3 PSR Testing with Conditioned Odor Cues Outside the Operant Environment

This version of cue exposure has the animal experience their assigned odor cue in a second neutral non-drug-paired environment. This technique divorces conditioned cues from drug-paired context and the performance of its drug self-administration associated behaviors (operant responding). The **(1)** operant training (10 weeks), **(2)** extinction training (1 week), **(3)** exposure to neutral odor, and **(4)** home cage rest are the same as in the previous cue exposure methods (*see* Fig. 6; Created with BioRender.com). However, for the PSR Test session, **(5)** the animals are randomly assigned to pretreatment groups (CS^+, CS^0, or CS^-) which occurs

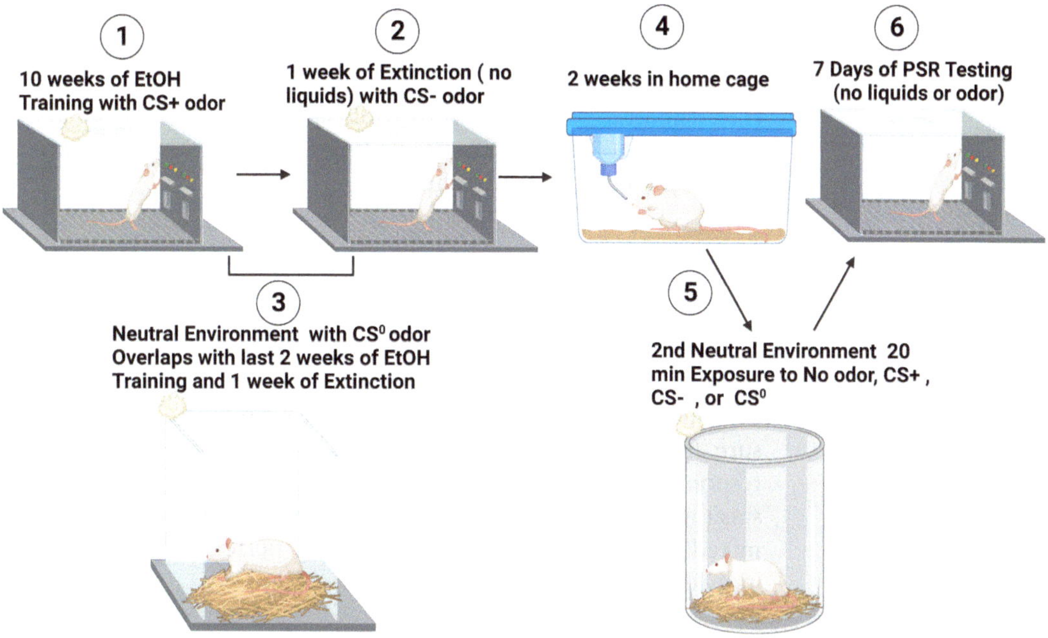

Fig. 6 Timeline for testing odor cues outside the operant environment

in a second neutral non-drug-paired environment [43, 67]. The second neutral environment is in a different shaped holding cage than the first neutral environment. Before each PSR test session, six drops of the pretreatment assigned odor (CS^+, CS^0, or CS^-) are administered into the bedding inside the cylinders; the rats are transferred into the cylinders and exposed to the assigned conditioned cue for 20 min (Hauser et al. [43, 67]). The used bedding inside the cylinders is disposed of and removed from the room, the cylinder is cleaned, and new bedding is added to the cylinder in between odors. If the second neutral environment comes with a wire mesh cage top, then the assigned condition stimulus (CS^+, CS^0, or CS^-) is emitted from a treated cotton ball (2–3 drops of the odor) on top of the cage. The cage tops are cleaned, the used bedding disposed of and removed from the room, and new bedding is added in between odors. **(6)** Immediately after the 20 min odor exposure, the rat is placed into the operant chamber (no odors or liquids in the operant chamber) for the 60 min PSR test session (*see* Fig. 6).

Our previous work demonstrated that pretreatment with a CS^+ in the non-drug paired environment can enhance the actions of an excitatory condition stimulus (CS^+; EtOH cue) on PSR-responding [43, 67]. Furthermore, the effects of a conditioned stimulus cue (CS^+; alcohol cue and CS^-; extinction cue) on PSR-responding were also observed up to 4 h after exposure to the cue [43]. These findings suggested that persistent heightened levels of EtOH-seeking caused by exposure to the CS^+ may create a mental state that gradually wears down the resistance of the individual to abstain from EtOH [43]. Contrarily, the ability of the CS^- to persistently inhibit PSR-responding may facilitate the potential therapeutic value of CS^- therapy/treatment [43]. In addition, CS^- exposure in the second non-drug paired (i.e., "neutral") environment prophylactically inhibited the ability of the CS^+ to evoke PSR-responding/EtOH-seeking behavior in the drug-paired environment [43].

6 Statistical Analyses

Overall operant responding (60 min) data are analyzed with a mixed factorial ANOVA with a between factor of conditioned cue exposure and a repeated measure of "session." The baseline measure for the factor of "PSR-session" is the average number of responses on the EtOH-paired lever for the last three extinction sessions. When significant effects ($p < 0.05$) are observed using the mixed ANOVA's, post hoc tests will be performed to determine individual group differences.

7 Conclusion

The PSR technique, in the context of EtOH-seeking behavior, has the benefit of assessing seeking behavior in the absence of a reinforcer and that PSR-responding persists across days following extinction and home cage rest. Therefore, the use of the PSR test may resemble the events that occur in patients with AUDs that have undergone detoxification in a clinical environment and, after a period of abstinence, may still have a strong desire to drink, with real-world environmental cues augmenting that desire to drink. Overall, the PSR experimental approach is an important tool for determining mechanisms underlying EtOH craving and relapse, which could lead to the development of novel pharmacotherapies to treat AUDs.

Acknowledgements

Preparation of this manuscript and its associated experiments were supported in part by AA020908 (ZAR), AA013522 (RLB), AA022287, AA007611, AA007462, and AA010721. The views expressed in this manuscript do not necessarily represent those of the National Institutes of Health (NIH) or the National Institute on Alcohol Abuse and Alcoholism (NIAAA). The products mentioned in this manuscript do not represent an endorsement of the particular manufacturer, but only reflect the products used by our laboratories. All authors claim no perceived or real conflicts of interest.

References

1. Ferri M, Amato L, Davoli M (2006) Alcoholics anonymous and other 12-step programmes for alcohol dependence. Cochrane Database Syst Rev 3:CD005032
2. Gawin FH, Kleber HD (1986) Abstinence symptomatology and psychiatric diagnosis in cocaine abusers. Clinical observations. Arch Gen Psychiatry 43:107–113
3. O'Brien CP, Childress AR, Ehrman R, Robbins SJ (1998) Conditioning factors in drug abuse: can they explain compulsion? J Psychopharmacol 12(1):15–22
4. Li P, Wu P, Xin X, Fan YL, Wang GB, Wang F, Ma MY, Xue MM, Luo YX, Yang FD, Bao YP, Shi J, Sun HQ, Lu L (2015) Incubation of alcohol craving during abstinence in patients with alcohol dependence. Addict Biol 20:513–522
5. Gordon SM, Sterling R, Siatkowski C, Raively K, Weinstein S, Hill PC (2006) Inpatient desire to drink as a predictor of relapse to alcohol use following treatment. Am J Addict 15:242–245
6. Pavlov IP (1927) Conditioned reflexes (G.V. Anrep, trans). Oxford University Press, London
7. Rodd ZA, Bell RL, Sable HJ, Murphy JM, McBride WJ (2004) Recent advances in animal models of alcohol craving and relapse. Pharmacol Biochem Behav 79(3):439–450
8. Martin-Fardon R, Weiss F (2013) Modeling relapse in animals. Curr Top Behav Neurosci 13:403–432
9. DiCiano P, Everitt BJ (2002) Reinstatement and spontaneous recovery of cocaine-seeking following extinction and different durations of withdrawal. Behav Pharmacol 13:397–405
10. Widholm JJ (2010) Extinction learning as a model of drug treatment and relapse: a behavioral overview. Open Addict J 3:57–62
11. Franken IH, de Haan HA, van der Meer CW, Haffmans PM, Hendriks VM (1999) Cue reactivity and effects of cue exposure in abstinent posttreatment drug users. J Subst Abus Treat 16(1):81–85
12. Cunningham CL, Fidler TL, Hill KG (2000) Animal models of alcohol's motivational effects. Alcohol Res Health 24(2):85–92
13. Epstein DH, Preston KL, Stewart J, Shaham Y (2006) Toward a model of drug relapse: an assessment of the validity of the reinstatement procedure. Psychopharmacology 189(1):1–16
14. Leeman RF, Heilig M, Cunningham CL, Stephens DN, Duka T, O'Malley SS (2010) Ethanol consumption: how should we measure it? Achieving consilience between human and animal phenotypes. Addict Biol 15(2):109–124
15. Samson HH (2000) The microstructure of ethanol drinking: genetic and behavioral factors in control of drinking patterns. Addiction 95(Suppl 2):S61–S72
16. Samson HH, Czachowski CL (2003) Behavioral measures of alcohol self-administration and intake control: rodent models. Int Rev Neurobiol 54:107–143
17. Lê AD, Funk D, Harding S, Juzytsch W, Fletcher PJ, Shaham Y (2006) Effects of dexfenfluramine and 5-HT3 receptor antagonists on stress-induced reinstatement of alcohol seeking in rats. Psychopharmacology 186(1): 82–92
18. de Wit H, Stewart J (1981) Reinstatement of cocaine-reinforced responding in the rat. Psychopharmacology 75(2):134–143
19. de Wit H, Stewart J (1983) Drug reinstatement of heroin-reinforced responding in the rat. Psychopharmacology 79(1):29–31
20. Shaham Y, Adamson LK, Grocki S, Corrigall WA (1997) Reinstatement and spontaneous recovery of nicotine seeking in rats. Psychopharmacology 130(4):396–403
21. Lê AD, Quan B, Juzytch W, Fletcher PJ, Joharchi N, Shaham Y (1998) Reinstatement of alcohol-seeking by priming injections of alcohol and exposure to stress in rats. Psychopharmacology 135(2):169–174
22. Richards JK, Simms JA, Steensland P, Taha SA, Borgland SL, Bonci A, Bartlett SE (2008) Inhibition of orexin-1/hypocretin-1 receptors inhibits yohimbine-induced reinstatement of ethanol and sucrose seeking in Long-Evans rats. Psychopharmacology 199(1):109–117
23. Kleber H, Gawin F (1986) Psychopharmacological trials in cocaine abuse treatment. Am J Drug Alcohol Abuse 12(3):235–246
24. Pickens CL, Airavaara M, Theberge F, Fanous S, Hope BT, Shaham Y (2011) Neurobiology of the incubation of drug craving. Trends Neurosci 34(8):411–420
25. Bouton ME, Bolles RC (1979) Role of conditioned contextual stimuli in reinstatement of extinguished fear. J Exp Psychol Anim Behav Process 5(4):368–378
26. Hamlin AS, Newby J, McNally GP (2007) The neural correlates and role of D1 dopamine receptors in renewal of extinguished alcohol-seeking. Neuroscience 146(2):525–536
27. Hamlin AS, Clemens KJ, McNally GP (2008) Renewal of extinguished cocaine-seeking. Neuroscience 151(3):659–670

28. Chaudhri N, Sahuque LL, Janak PH (2009) Ethanol seeking triggered by environmental context is attenuated by blocking dopamine D1 receptors in the nucleus accumbens core and shell in rats. Psychopharmacology 207: 303–314
29. Chaudhri N, Sahuque LL, Schairer WW, Janak PH (2010) Separable roles of the nucleus accumbens core and shell in context- and cue-induced alcohol-seeking. Neuropsychopharmacology 35:783–791
30. Katner SN, Magalong JG, Weiss F (1999) Reinstatement of alcohol-seeking behavior by drug-associated discriminative stimuli after prolonged extinction in the rat. Neuropsychopharmacology 20(5):471–479
31. Katner SN, Weiss F (1999) Ethanol-associated olfactory stimuli reinstate ethanol-seeking behavior after extinction and modify extracellular dopamine levels in the nucleus accumbens. Alcohol Clin Exp Res 23(11): 1751–1760
32. Macintosh JJ (1977) Stimulus control: attentional factors. In: Honig WK, Staddon JER (eds) Handbook on Operant Behavior. Prentice-Hall, Englewood Cliffs, pp 162–241
33. Robbins SJ (1990) Mechanisms underlying spontaneous recovery in authoshaping. J Exp Psychol Anim Behav Processes 16:235–249
34. Rodd-Henricks ZA, Bell RL, Kuc KA, Murphy JM, McBride WJ, Lumeng L, Li TK (2002) Effects of ethanol exposure on subsequent acquisition and extinction of ethanol self-administration and expression of alcohol-seeking behavior in adult alcohol-preferring (P) rats: I. Periadolescent exposure. Alcohol Clin Exp Res 26(11):1632–1641
35. Rodd-Henricks ZA, Bell RL, Kuc KA, Murphy JM, McBride WJ, Lumeng L, Li TK (2002) Effects of ethanol exposure on subsequent acquisition and extinction of ethanol self-administration and expression of alcohol-seeking behavior in adult alcohol- preferring (P) rats: II. Adult exposure. Alcohol Clin Exp Res 26(11):1642–1652
36. Getachew B, Hauser SR, Dhaher R, Katner SN, Bell RL, Oster SM, McBride WJ, Rodd ZA (2011) CB1 receptors regulate alcohol-seeking behavior and alcohol self-administration of alcohol-preferring (P) rats. Pharmacol Biochem Behav 97(4):669–675
37. Dhaher R, Hauser SR, Getachew B, Bell RL, McBride WJ, McKinzie DL, Rodd ZA (2010) The Orexin-1 receptor antagonist SB-334867 reduces alcohol relapse drinking, but not alcohol-seeking, in alcohol-preferring (P) rats. J Addict Med 4(3):153–159
38. Dhaher R, Toalston JE, Hauser SR, Bell RL, McKinzie DL, McBride WJ, Rodd ZA (2012) Effects of naltrexone and LY255582 on ethanol maintenance, seeking, and relapse responding by alcohol-preferring (P) rats. Alcohol 46(1):17–27
39. Knight CP, Hauser SR, Deehan GA Jr, Toalston JE, McBride WJ, Rodd ZA (2016) Oral conditioned cues can enhance or inhibit ethanol (EtOH)-seeking and EtOH-relapse drinking by alcohol-preferring (P) rats. Alcohol Clin Exp Res 40(4):906–915
40. Hauser SR, Ding ZM, Getachew B, Toalston JE, Oster SM, McBride WJ, Rodd ZA (2011) The posterior ventral tegmental area mediates alcohol-seeking behavior in alcohol-preferring rats. J Pharmacol Exp Ther 336(3):857–865
41. Hauser SR, Deehan GA Jr, Toalston JE, Bell RL, McBride WJ, Rodd ZA (2014) Enhance alcohol-seeking behavior by nicotine in the posterior ventral tegmental area of female alcohol-preferring (P) rats: modulation by serotonin-3 and nicotinic cholinergic receptors. Psychopharmacology 231(18): 3745–3755
42. Hauser SR, Deehan GA Jr, Dhaher R, Knight CP, Wilden JA, McBride WJ, Rodd ZA (2015) D1 receptors in the nucleus accumbens-shell, but not the core, are involved in mediating ethanol-seeking behavior of alcohol-preferring (P) rats. Neuroscience 295:243–251
43. Hauser SR, Deehan GA Jr, Knight CP, Waeiss RA, Truitt WA, Johnson PL, Bell RL, McBride WJ, Rodd ZA (2019) Conditioned stimuli affect ethanol-seeking by female alcohol-preferring (P) rats: the role of repeated-deprivations, cue-pretreatment, and cue-temporal intervals. Psychopharmacology 236(9):2835–2846
44. LeCocq MR, Lahlou S, Chahine M, Padillo LN, Chaudhri N (2018) Modeling relapse to Pavlovian alcohol-seeking in rats using reinstatement and spontaneous recovery paradigms. Alcohol Clin Exp Res 42(9): 1795–1806
45. Anton RF (1999) What is craving? Models and implications for treatment. Alcohol Res Health 23(3):165–173
46. Rodd ZA, Anstrom KK, Knapp DJ, Racz I, Zimmer A, Serra S, Bell RL, Woodward DJ, Breese GR, Colombo G (2005) Factors mediating alcohol craving and relapse: stress, compulsivity, and genetics. Alcohol Clin Exp Res 29(7):1325–1333
47. Hauser SR, Deehan GA Jr, Knight CP, Toalston JE, McBride WJ, Rodd ZA (2016) Parameters of context-induced ethanol (EtOH)-seeking in alcohol-preferring (P) rats: temporal

analysis, effects of repeated deprivation, and EtOH priming injections. Alcohol Clin Exp Res 40(10):2229–2239

48. Rodd ZA, McKinzie DL, Bell RL, McQueen VK, Murphy JM, Schoepp DD, McBride WJ (2006) The metabotropic glutamate 2/3 receptor agonist LY404039 reduces alcohol-seeking but not alcohol self-administration in alcohol-preferring (P) rats. Behav Brain Res 171(2):207–215
49. Deehan GA Jr, McKinzie DL, Carroll FI, McBride WJ, Rodd ZA (2012) The long-lasting effects of JDTic, a kappa opioid receptor antagonist, on the expression of ethanol-seeking behavior and the relapse drinking of female alcohol-preferring (P) rats. Pharmacol Biochem Behav 101(4):581–587
50. Hauser SR, Waeiss RA, Molosh AI, Deehan GA Jr, Bell RL, McBride WJ, Rodd ZA (2020) Atrial natriuretic peptide (ANP): a novel mechanism for reducing ethanol consumption and seeking behaviors in female alcohol preferring (P) rats. Peptides 134:170403
51. Bell RL, Hauser S, Rodd ZA, Liang T, Sari Y, McClintick J, Rahman S, Engleman EA (2016) A genetic animal model of alcoholism for screening medications to treat addiction. Int Rev Neurobiol 126:179–261
52. Domi E, Domi A, Adermark L, Heilig M, Augier E (2021) Neurobiology of alcohol seeking behavior. J Neurochem 157(5):1585–1614
53. Funk D, Mak D, Coen K, Lê AD (2022) Role of alpha-2 adrenergic and kappa opioid receptors in the effects of alcohol gavage-induced dependence on alcohol seeking. Behav Brain Res 434:114032
54. Nelson BS, Fulenwider HD, Nennig SE, Smith BM, Sequeira MK, Chimberoff SH, Richie CT, Cheng K, Rice KC, Harvey BK, Heilig M, Schank JR (2019) Escalated alcohol self-administration and sensitivity to yohimbine-induced reinstatement in alcohol preferring rats: potential role of neurokinin-1 receptors in the amygdala. Neuroscience 413:77–85
55. Samson H, Files F, Brice G (1996) Patterns of ethanol consumption in a continuous access situation: the effect of adding a sweetener to the ethanol solution. Alcohol Clin Exp Res 20(1):101–109
56. Sanchis-Segura C, Spanagel R (2006) Behavioural assessment of drug reinforcement and addictive features in rodents: an overview. Addict Biol 11(1):2–38
57. Augier E, Flanigan M, Dulman RS, Pincus A, Schank JR, Rice KC, Kejun C, Heilig M, Tapocik JD (2014) Wistar rats acquire and maintain self-administration of 20% ethanol without water deprivation, saccharin/sucrose fading, or extended access training. Psychopharmacology 231(23):4561–4568
58. Hauser SR, Getachew B, Oster SM, Dhaher R, Ding ZM, Bell RL, McBride WJ, Rodd ZA (2012) Nicotine modulates alcohol-seeking and relapse by alcohol-preferring (P) rats in a time-dependent manner. Alcohol Clin Exp Res 36(1):43–54
59. Paxinos G, Watson C (2005) The rat brain in stereotaxic coordinates, 5th edn. Academic Press, New York
60. Rosenthal MZ, Kutlu MG (2014) Translation of associative learning models into extinction reminders delivered via mobile phones during cue exposure interventions for substance use. Psychol Addict Behav 28(3):863–871
61. Willcocks AL, McNally GP (2014) An extinction retrieval cue attenuates renewal but not reacquisition of alcohol seeking. Behav Neurosci 128:83–91
62. McNally GP (2014) Extinction of drug seeking: neural circuits and approaches to augmentation. Neuropharmacology 76 (Pt B):528–532
63. Rich MT, Abbott TB, Chung L, Gulcicek EE, Stone KL, Colangelo CM, Lam TT, Nairn AC, Taylor JR, Torregrossa MM (2016) Phosphoproteomic analysis reveals a novel mechanism of CaMKIIα regulation inversely induced by cocaine memory extinction versus reconsolidation. J Neurosci 36:7613–7627
64. Rich MT, Huang YH, Torregrossa MM (2019) Plasticity at Thalamo-amygdala synapses regulates cocaine-cue memory formation and extinction. Cell Rep 26:1010–1020.e5
65. Torregrossa MM, Gordon J, Taylor JR (2013) Double dissociation between the anterior cingulate cortex and nucleus accumbens core in encoding the context versus the content of pavlovian cocaine cue extinction. J Neurosci 33:8370–8377
66. Bender BN, Torregrossa MM (2021) Dorsolateral striatum dopamine-dependent cocaine seeking is resistant to Pavlovian cue extinction in male and female rats. Neuropharmacology 182:108403
67. Hauser SR, Deehan GA Jr, Knight CP, Waeiss RA, Engleman EA, Ding ZM, Johnson PL, McBride WJ, Truitt WA, Rodd ZA (2023) Inhibitory and excitatory alcohol-seeking cues distinct roles in behavior neurochemistry and mesolimbic pathway in alcohol preferring (P) rats. Drug and Alcohol Dependence 246: 109858

Chapter 3

The Delay Discounting Procedure: Methodology and Flexibility

Meredith Halcomb

Abstract

Delay discounting (DD) is a well-established method for evaluating behavioral impulsivity. It has been implemented extensively over the past few decades to identify associations between impulsivity and numerous psychopathologies. A valuable aspect of the DD task is its flexibility, allowing more targeted investigations of specific populations or disorders and several methodological factors may be manipulated by the researcher to maximize its utility. This chapter describes the precise methods through which our lab administers the DD task, including computer setup, instructions for participants, and a step-by-step explanation of trial processes which generate indifference points across several delays. Several different elements of the task are also considered, including the use of real or hypothetical rewards, the type of task used, and framing of questions. Each of these factors plays a significant role in the type of data acquired, the population best evaluated, and how it should be interpreted.

Key words Delay discounting, Adjusting amounts, Real rewards, Hypothetical rewards, Framing, Methods

1 Introduction

Impulsivity, as a construct, generally refers to the tendency to engage in action without forethought or regard to consequences. As impulsive behavior is a hallmark of numerous psychopathologies, researchers have developed and refined several methods to evaluate and manipulate both state and trait expressions of impulsivity. One such method, the delay discounting (DD) task, is based upon the premise that the subjective value of some reward diminishes as the time to receipt increases [1]. For example, if a participant is given a choice between $10 today or $50 a week from now, they may choose $50 a week from now; however, if the choice is between $10 today or $50 a year from now, that preference may reverse. This task is posited to emulate everyday life decisions, such as: *Have another drink and feel good tonight* or *Go home now and feel good tomorrow*. Preferential, consistent choice of the more

Melissa A. Cyders (ed.), *Translational Research Methods for Alcohol Use Disorders*, Neuromethods, vol. 201,
https://doi.org/10.1007/978-1-0716-3267-3_3,

immediate, smaller reward over the delayed, larger reward is indicative of suboptimal choice, leading to less reward over time, and considered an indication of impulsive behavior.

Steep delay discounting (greater impulsivity) is linked to numerous behavioral and psychological disorders. There is evidence that delay discounting is associated with gambling [2], risky sexual behavior [3], obesity and binge eating disorder [4], and other health-risk behaviors [5], including compliance with COVID-19 protective measures [6]. Currently drinking alcohol use disorder (AUD) participants discount rewards more steeply than abstinent AUD participants and social drinking controls [7], while cigarette smokers discount monetary rewards more steeply than either ex-smokers or matched controls [8]. Recent research also indicates a negative relationship between DD behavior and length of recovery before relapse in AUD [9], and discounting rates are strong indicators of cigarette smoking relapse [10]. Greater discounting is observed in both cocaine- and methamphetamine-dependent individuals [11, 12], as well as heroin use [13]. The obvious implication is that the ability to measure DD reliably is a valuable mechanism for researchers seeking greater understanding of psychopathologies and for those developing possible interventions.

Fortunately, the DD task has good test-retest reliability and stability [14, 15] and shows moderate levels of heritability in a longitudinal twin study [16]. It also demonstrates limited practice effects and is sensitive enough to detect within-subject changes [17]. Finally, DD is often administered in pre-clinical studies, lending it the flexibility to assess pharmacological or genetic manipulations and good translational capabilities. A quick PubMed search of delay discounting yields results from over 2000 studies just within the past 10 years, attesting to the fact that the DD task has evolved into a prodigious research tool. However, this explosion of utilization is accompanied by greater deconstruction of variable aspects of the task (such as framing and context), differences in administration (including hypothetical versus real rewards or adjusting amount versus adjusting delay), and even a more critical view of the overall concept of DD and its identification as a "core" process underlying behavior abnormalities [18].

In this chapter, I will describe the DD task implemented in our studies through a step-by-step process, including information on the participant instructions, as well as calculation of preferred outcome measures. In addition, numerous variations available and considerations for administration are described. While the goal of this chapter is not to be a thorough review of the DD task, it will provide a roadmap for administration, in conjunction with alternative possibilities for specific task parameters, such as task and reward type, framing effects, and context-dependent effects.

2 Materials

For the methods described here, the primary tool will be a laptop or desktop computer, with mouse, programmed with E-Prime software (Psychology Software Tools, Inc., Sharpsville, PA). Although E-Prime is the most common software used for DD, PsychoPy Software (Peirce, 2007) is also available. Alternatively, Mechanical Turk (MTurk) by Amazon is a simple, easy method of administering the task to large samples of people, with subjects completing the task online for a specific compensation. It should be noted that the computerized version described herein requires a programmer to develop the computer code, which is not provided.

3 Methods

1. *Task Setup*

   E-Prime dongle is inserted into the USB port prior to laptop boot-up. Once E-Prime is opened, follow these steps:

   (a) File → Open → Delay Discounting Task.

   (b) Select the task to load. Our study has four versions that are alternated across subjects.

   (c) Once task is loaded: Select the "running man" icon in task bar.

   (d) E-Run pop-up asks for subject's name. Our study assigns a number for each subject, which is the number entered here. Number cannot exceed 30,000.

   (e) Next E-Run pop-up asks for session number. Our subjects only complete 1 session, so put in 1.

   (f) Program then asks to confirm subject and session numbers. At this point, these can both still be changed.

   (g) Next screen includes instructions. Mouse click takes screen to sample trial (*see* below).

2. *The Administration of the Procedure*

   The task is administered in a small, quiet room without distractions. Once instructions are given and the participant is given a brief demonstration of the task, investigator leaves the room to allow privacy and ensure that their presence does not influence responding. Once the laptop is set up, the participant is given the following instructions:

   > Today you will be making a choice about money that can be delivered either today or in the future. The task takes about 20 min to complete. Please make your choices deliberately and carefully, but without wasting time. You should make every choice as though it is real, because one of your choices will be real. One choice will be delivered at random by the computer and paid out according to your choice of amount and delay.

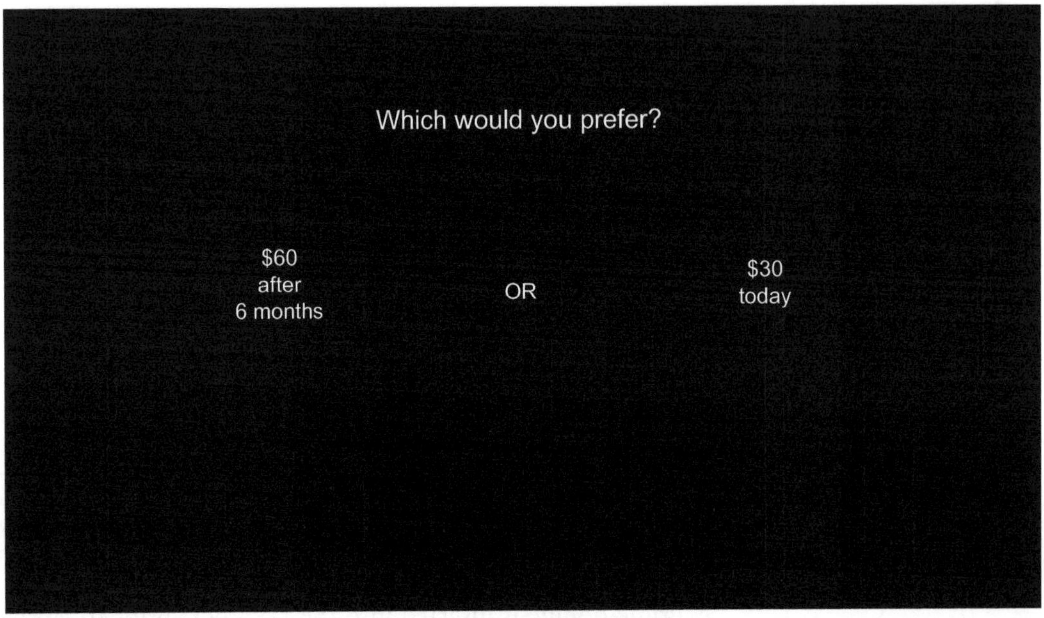

Fig. 1 Example trial administered and explained to each subject prior to testing administration

> On the computer screen, you will be faced with two options for gaining money. One of them will result in payment today, and one of them will result in future payment. You can choose either one. If you want to select the choice on the left, press the left mouse button. If you want the choice on the right, select the right mouse button. Note that the choices will change sides, meaning that a delayed choice can be on the right side for one trial, but for the next trial, it can be on the left side. Between every choice, there will be a pause, and then another choice will be displayed.
>
> We will begin with an example so that you see what the choices look like.
>
> *Go through example.*
>
> Do you have any questions?

The subject is walked through the example (*see* Fig. 1) prior to the initiation of the task. To avoid possible framing issues or reference points (*see* below), subjects are administered one of four different versions of the task with differing initial delays and amounts. Although this is intended as a preventative measure, there is no evidence that the version of the task completed is associated with significant differences in response patterns.

3. *The Adjusting-Amount (AA) Paradigm*

The standard amounts used for this task are $20 and $60, available after set delays of 2 days, 1 week, 1 month, 6 months, 1 year, and 5 years. The initial presentation includes one of the standard amounts at one of the delays and an amount available immediately (today) that is 1/2 of the delayed amount. The amount available immediately will titrate higher or lower, based upon participant's choices. This titration repeats after every choice to establish the specific indifference point for each delay and

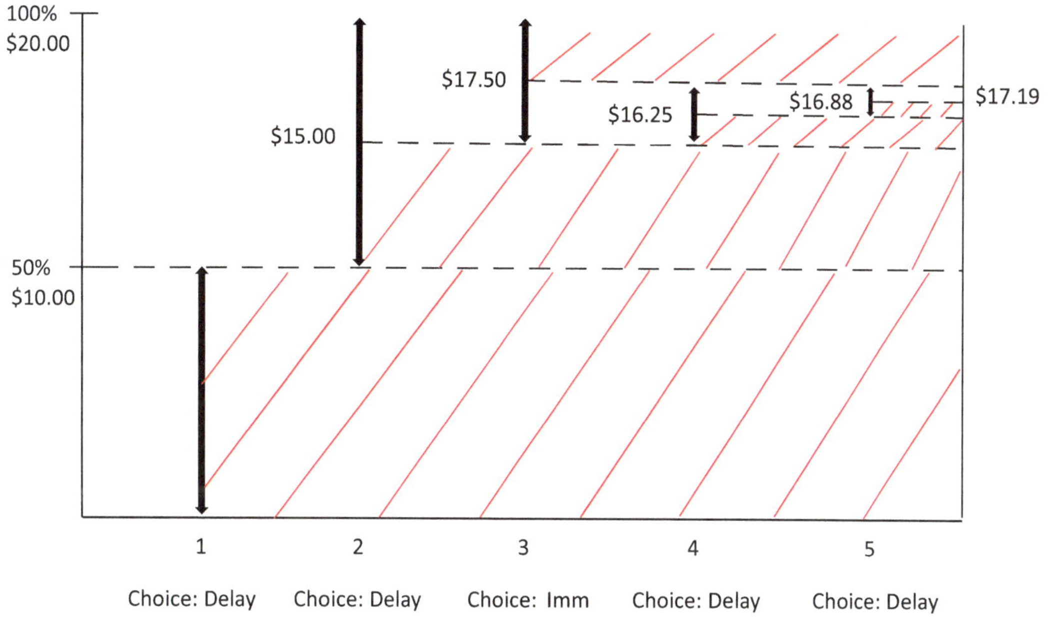

Fig. 2 Trial-by-trial example of how the program reaches an indifferent point. For each choice, the amount available at the immediate option is titrated higher or lower by 20% of the previous amount. *Imm* immediate

incorporates previous titration levels. The amount of adjustment for each choice is influenced by previous choices, wherein each increase or decrease in the immediate amount is 1/2 of the difference from the previous adjusted amounts. The following is an example of how five trials at a specific delay of 2 days reaches the indifference point (*see* Fig. 2). This is a hypothetical trial-by-trial example of what results from different choices made by the participant:

(a) Start of session: $10.00 available immediately or $20.00 after 2 days.

(b) Participant chooses Delayed amount.

(c) Since participant chooses the delayed amount over $10.00 immediately, this is now considered the lowest amount from which to titrate. It is assumed that the participant would not choose an immediate amount lower than $10. Therefore, the difference between the two choices in the next trial is now $10.00 and 20.00. The amount available immediately increases by 1/2 that difference ($5.00).

(d) Trial 2: $15.00 available immediately or $20.00 after 2 days.

(e) Participant chooses delayed amount.

(f) Step c repeats. The lowest amount is now $15.00, making the base difference between $15.00 and $20.00. The immediate amount titrates up again by 1/2 of the difference ($2.50).

(g) Trial 3: $17.50 available immediately or $20.00 after 2 days.

(h) Participant chooses the immediate amount.

(i) Choice of the immediate amount indicates that the subjective value of $20.00 after 2 days is less than $17.50. Therefore, the base difference for the immediate choice is between $15.00 and $17.50. The amount available immediately will titrate down by 1/2 of that difference ($1.25).

(j) Trial 4: $16.25 available immediately or $20.00 after 2 days.

(k) Participant chooses delayed amount.

(l) Step c repeats. The lowest amount available immediately is now $16.25, making the base difference to be between $16.25 and $17.50 (upper limit has not changed). The immediate amount will titrate up by 1/2 of this difference ($0.63).

(m) Trial 5: $16.88 available immediately or $20.00 after 2 days.

(n) Participant chooses the delayed amount.

(o) This choice results in a final titration to the indifference point for this delay. The difference between $16.88 and $17.50 is $0.62. Immediate amount is calculated by increasing by 1/2 of this amount ($0.31). The final indifference point for this delay is therefore $17.19.

These steps are completed for each delay. Trials are presented in a randomized order so participants may complete a trial for 2 days of delay, followed by a trial for 6 months of delay, followed by a trial for 1 week of delay, etc., with five trials for each delay for each of the standard amounts ($20.00 and $60.00) for a total of 60 trials.

Presentation of immediate or delayed choices are counterbalanced so that on 1/2 of the trials the immediate choice is on the left and for the other 1/2 of the trials, the immediate choice is on the right. This technique aids in preventing the influence of reference points (*see* below). Respondents are given unlimited time to complete each choice, but may not continue to the next trial without selecting an alternative. Once a choice is completed, the screen briefly changes to black before the next trial is presented.

4. *Data analysis*

The adjusting amount paradigm generates indifference points across several different delays. For analysis, we calculate both *k* values and area under the curve (AUC), two separate measures of analyzing discount rates. The indifference point can be described as the point at which the adjusted amount is subjectively equal to the delayed amount. Since our lab uses both types of analysis, we will discuss both here, rather than in a separate section.

The most generally accepted function for calculation of discounting is the hyperbolic equation:

$$V = A/(1 + kD)$$

where V is the time-discounted value of the reward, A is the subjective present value (the adjusted amount), D is the delay time, and k is the discounting coefficient (the rate at which subjective value decreases). The constant "1" is added to the equation to prevent the possibility of the curve extending to infinity at short intervals. Although this is a common technique for assessing discounting curves, there are some potential issues with this method that should be acknowledged. First, k is only appropriate when the curve can be accurately described by hyperbolic decay. To that end, researchers analyze data for violations of specific criteria for it to be considered systematic and viable [19]. A violation occurs if participants do not decrease preference for the delayed reward as that delay becomes larger. A second violation occurs if the participant does not discount the longest delayed reward by at least 10% more than the shortest delayed reward. If either of these violations happen, the data is considered nonsystematic and discarded. It has been suggested that researchers are overly focused on these modeling criteria and that discarding data which does not conform to the hyperbolic function leads to over-confirmation of the model's utility [18].

Additionally, because there is no upper or lower bound on the parameter, a second problem with k is the tendency for skewness in the dataset, violating normalcy assumptions, and preventing use of parametric testing [20]. Values generally need log transformation to normalize them for analysis. Finally, although there is evidence of alternate-form test-retest reliability (Kirby monetary choice task and adjusting amount task [21]), k values can be highly variable between different commodities, such as food or drug [22].

An alternative approach to delay discounting analysis is AUC, which requires no assumptions about the distribution of data. It is calculated by plotting the coordinates of indifference points at each delay and then computing the area of the trapezoids created. The equation for the AUC is:

$$x_1 - x_2 \left[(y_1 + y_2)/2 \right]$$

where the ordinate represents the proportion of the non-discounted value (delays), and the abscissa represents the proportion of maximum possible delay (indifference points for those delays) [20]. The resulting values are summed and will be in the range from 0.0 to 1.0, wherein the steeper the discounting curve, the smaller the AUC value. This upper and lower boundary allows parametric statistical testing, since it limits the skewness observed in k values. Conversely, normalization can also be a disadvantage since it prevents comparison between experiments with different delays and monetary values without adjusting for those differences. In addition, two AUC discounting functions may be the same while describing two different curves. For example, if participant 1 discounts sharply at the first delays, then levels off, they may have the same AUC as participant 2, who discounts more slowly overall.

Early DD research posited that discounting occurred in an exponential manner; however, hyperbolic models consistently provide a better fit for DD data. While this hyperbolic discounting is most found in the literature, it should be noted that a study using "fuzzy" outcome measures (handful of quarters) versus "concrete" outcomes noted that an exponential curve better describes discounting for clear delayed monetary rewards or food rewards for a large subset of the sample [23]. Other studies use a hyperboloid function for assessment, adding the exponent *s*, which represents the psychophysical scaling of amount, delay, and probability [20]. Best practice is to analyze raw data, look at discounting rates derived from both formulas, and determine which method is a best-fit for the data.

4 Notes

The amount of research evaluating the various aspects of the DD task, including everything from reward type to question presentation to intertrial interval, is commensurate with the prolific application of the task within the past two decades. To ensure the task is valid and reliable, numerous investigations have sought to identify potential issues and advance superior methods. The results of several of these studies are described below, along with occasional recommendations from authors for best practices in administration. It should be noted that although the majority of the data discussed are generated through the computer-administered version, there is also a paper version available [24], which presents a fixed set of choices between smaller, immediate rewards and larger, delayed rewards. While this questionnaire (and other, similar questionnaires) provides a simple and quick estimate of discounting, it may not be as sensitive as a computerized AA version. The questionnaires produce discrete *k* values for each delay option, while the adjusting methods allow for generation of a continuous *k* value across all delays. A comparison between these two methods found that they could not be used interchangeably [21]. While the questionnaire is still implemented occasionally, this chapter will focus primarily upon methods associated with computer administration.

Real Versus Hypothetical Rewards

In presenting the DD task, the first question is what type of reward will be available. For monetary DD tasks (and also consumptive tasks, such as food or alcohol), the reward can be real or hypothetical. The use of hypothetical rewards allows greater breadth of dollar amounts and delay times. For example, while the hypothetical method may provide a choice between $100 today or $1000 ten years from now, that question would not be feasible using real rewards. Conversely, the use of real rewards, while not as flexible in reward value or time to receipt, may better reflect real-world decision-making. In addition, it may be inadvisable to present certain populations with monetary rewards which can be used to purchase illicit substances. Investigators who utilize hypothetical rewards acknowledge that choices made may not reflect choices with real outcomes; however, the discounting curves typically derived from these studies are hyperbolic in nature, which provides the best fit for discounting data [25].

This question is important because the literature shows that measures of other behaviors may vary based upon reward type. For instance, there is evidence of decreased risk taking when paired with real consequences [26], variables which influence risk taking (positive feelings) of real rewards have no effect on hypothetical risk taking [27], real monetary rewards are stronger incentives than hypothetical rewards [28], and altruistic behavior is altered, wherein people are more likely to donate higher amounts of money in hypothetical situations [29]. Extensive research evaluating differences in discounting in real versus hypothetical rewards has led to the tentative conclusion that both generate similar hyperbolic curves. Early research directly comparing the two using within-subject designs found no significant differences in degree of hyperbolic discounting or measurements of area under the curve [30, 31]. Although promising, it is possible that participants were responding to the second administration of the task exactly as they had the first administration, therefore confounding results. For example, if the first task was available with real rewards, it may have influenced subjects to respond a similar way in the hypothetical reward paradigm. Further research implemented a between-subject comparison, wherein participants received similar instructions up until the point at which their actual reward was outlined. Subjects in the hypothetical condition were informed they would be compensated with $5.50, while participants in the real condition were told that one choice would be chosen at random [32]. This study also found no significant effect of reward type. These results are replicated in a more recent study varying the hypothetical rewards extensively using a prisoners-dilemma type game [33].

Research from neuroimaging studies is not as straightforward. Early work in which task administration was conducted in the scanner found no differences in discounting between real and hypothetical rewards [34]. This study also found that both reward versions of DD were associated with activation in regions previously identified to be integral to the decision-making process, including areas involved in executive function. In contrast, a study using real versus hypothetical rewards of actual objects found that real rewards were associated with greater activation in orbitofrontal cortex (OFC) and ventral striatum [35]. Although this suggests that there may be alternate systems assessing value of different rewards, it should be noted that this was not a DD task, because all choices were presented as being available immediately.

Type of Task Administration

The decision to implement real or hypothetical rewards will likely influence the methodology of the task administration itself. One important aspect of how to present the DD task is the research question itself. A much-queried element of temporal discounting is whether DD measures impulsivity that is "state"- or "trait"-based and different versions of DD may better reflect assessment of one over the other. Converging evidence from self-report studies suggests that impulsivity is a trait which can be measured [36, 37] and the DD task demonstrates good test-retest reliability, beginning in adolescence [15, 38]. Research implementing pharmacological manipulations to alter DD responding is less clear. For instance, acute alcohol administration may increase impulsivity [39], decrease impulsivity [40], or have no effect [41]. A possible underlying factor contributing to these findings is the type of DD task administered. Given the immense potential for variability in DD presentation (standard amounts, delay length, etc.), this chapter will focus on broad methodologies currently in use. The type of task described above is the adjusting amount procedure based upon Du et al. [42], and while it is ideal for our research, there are other options which may be better suited to other types of research.

1. Researchers may establish indifference points without response-based adjustment of the immediate reward using a random order survey. While the adjusting amounts procedure provides a basis through which the participant actively influences what the amount of the immediate reward can be, a more straightforward version of this is available. In this instance, there are standard amounts for both the immediate and delayed choices presented at several different delays in a randomized order [43]. This method has the benefit of availability in both computerized and pencil and paper formats, since it does not require a program to alter amounts. When directly compared to the adjusting amounts task described above, there was no significant difference in discounting as measured by either the hyperboloid or AUC function [44].
2. An alternative, but similar, type of administration to the adjusting amount procedure is the adjusting delay procedure [45]. In this task, the amounts of the immediate and delayed rewards are held constant, but the delay to the larger, later reward (LLR) is titrated until each amount is likely to be chosen. As in the adjusting amount task, there are several initial delays associated with 4–6 choice trials each, presented in a randomized manner, allowing for calculation of indifference points.

3. A posited mechanism for determining state-related changes in impulsivity is the use of the experiential discounting task (EDT) [46]. This method is similar to those utilized in pre-clinical models of DD in which if the subject/participant chooses the delayed option, they must experience the actual delay during the task itself, rather than imagining the receipt of the reward at some future time. The authors suggest that traditional tasks may not distinguish state changes in discounting because they lack the robustness necessary to detect changes in moment-to-moment variability. They found that the EDT was able to identify differences in responding in sleep-deprived versus non-sleep-deprived states and rates were best characterized by the hyperbolic function typically used to described DD data. This task uses smaller dollar amounts as reward and is being implemented across several different domains; however, it should be noted that there is evidence that the EDT is associated with steeper discounting rates than alternant methods of DD administration [47]. This may contribute to the increased impulsive behavior under acute alcohol administration observed previously [39]. In addition, although the lower denominations required for reward in this study make it cost-attractive, it may not be sufficient to characterize responding for larger rewards and time spans. For instance, one of the hallmarks of addiction is action without forethought of long-term consequences, which the EDT may not be able to capture.

   Implementation of this task across several commodities provides researchers with further avenues of investigation. For instance, there is evidence that there is a distinct difference in discounting associated with postponing a delayed reward (used in a traditional DD task) and having to "wait" for a reward [48–50]. One possible mechanism underlying this disassociation is time perception. Recent work suggests that those who more steeply discount future rewards also perceive time as passing more slowly when forced to wait during that delayed period [51]. Another study found that participants who were in the topmost and lowest extremes in discounting showed higher variability in time perception accuracy, which was also positively correlated with alcohol use [52]. In the EDT task, subjects must experience "waiting". This comprises two separate aspects: direct costs (boredom) and opportunity costs (lack of ability to pursue alternate things). The direct cost may influence responding as the session continues, as boredom compounds, resulting in changing response patterns and steeper discounting toward the end [53]. Understanding possible influences on DD during the EDT and similar tasks is important for both task setup and interpretation and it should be remembered that results from the EDT and other discounting methods are not always correlated [54].

4. Although the advent of the adjusting amount and adjusting delay paradigms allows researchers to ascertain indifference points in a relatively short time frame, their length may still prove prohibitive. To that end, a 5-trial adjusting delay discounting task, which can generate discount rates in less than a minute, may be more appropriate when time is limited, or participants are particularly low in attentional abilities [55]. Rather than determining indifference points for each participant, the 5-trial task calculates the Effective Delay 50% (ED_{50}), the inverse of the discount rate described by *k*. In this task, the reward amounts remain stable (some specific amount for the delayed reward and 1/2 that amount for the immediate reward), while the delay to the larger reward is adjusted across 5 trials, based upon participant choices. While this task may be well-suited for studies with restricted lab time, it is not ideal for research seeking discounting curves, since it does not generate a series of indifference points. Additionally, the generation of indifference curves allows researchers to exclude data which violates fundamental assumptions of logical responding in the task, which can add variability.
5. Other types of tasks are available which are typically variations of those listed above. These include the fixed alternative survey, which is similar in concept to the random order survey [56], and the Kirby monetary choice questionnaire [24], either of which can be administered without the use of a computer. The visual analog scale (VAS) [57] is similar to the EDT in that participants experience the delay to reward during the testing session; however, the delayed period is "framed" one of four ways. The framing can be "free", meaning the participant can leave the room during the entire delay; "return", in which the subject can leave the room, but must return periodically and check in during the delay; "browse", in which the subject has to stay in the room during the delay, but could scan the internet or their phones; or "wait" during which the subject has to remain at the computer without any other form of entertainment. Finally, in the fill-in-the-blank version of DD [58], participants are told to pretend they won the lottery and that they could receive either $500 now or some larger amount in the future. They are then asked to supply the amount of the delayed option that would make the two choices equally attractive. This is repeated with several "prize" amounts and delays.

There are several factors contributing to the type of task which should be administered, including the outcome measure desired (for instance, state versus trait level impulsivity), time constraints, funding, and population of interest. Although the versions described herein constitute the most common found in the literature, investigators are able to alter several parameters of the task to suit their own needs; therefore, this may not exemplify an exhaustive list of all possible renditions.

Framing

There are a myriad ways in which the slightest alteration in trial presentation may affect behavioral response. For instance, a study on gambling and choice found that presenting payoffs in columns instead of rows was sufficient to influence choice [59]. One of the earliest described factors in DD is the magnitude effect, wherein larger rewards are discounted less steeply than smaller rewards in both pre-clinical and clinical models [60, 61]. Magnitude is particularly important when evaluating AUD, as participants in this population are more sensitive to this effect [62]. This is a stable, robust aspect of DD and is an important consideration when designing a task. For instance, if the population under investigation is likely very impulsive, it may be more informative to provide a DD task with hypothetical rewards of very high amounts to prevent floor effects and allow for more variability in responding. Conversely, studies in healthy populations, which tend to discount less steeply, may benefit from lower monetary reward amounts to increase discounting.

There are several demographic factors that need to be addressed or, at the least, acknowledged, when designing a DD task. Although there is no consistent evidence for sex differences in DD [63, 64], there are observed associations with response to drug in pre-clinical models [65, 66], psychopathologies in clinical models [67, 68], and across cultures [69]. Sex should always be included as a covariate and be tested for main effects. Age of participants should also be considered in data analyses. Research demonstrates that children, young adults, and the elderly all discount more steeply than those in early/middle adulthood in western culture [70], while there is a steady increase in discounting across the life span in a sample from China [71]. Unfortunately, there are also null and inconsistent results, as well [72, 73]. While the current research may not provide a delineated timeline for changes in DD responding across the lifespan, there are notable changes in neural function which occur throughout life [74], so best practice would be to include age as covariate in data analyses, even when there are no significant differences between groups.

Finally, given the monetary nature of the DD task, the financial status of the participant and state of the economy are important factors to consider. Participants with lower income discount rewards more steeply [75] and high rates of inflation and interest rates can increase impulsive responding [76, 77]. In addition, asking participants to consider their current and future finances decreases discounting [78]. Lempert et al. [79] suggest that differences in financial stability affect the subjective value of money, so participants should be matched for socio-economic status (SES). Another possible avenue for investigation is to evaluate participant reward magnitude sensitivity and associated neuroeconomic influences [80].

4.1 Types of Framing That Reduce Discounting

Although there are several methods for decreasing discounting as part of the research or interventional design, the majority of those are beyond the scope of this chapter, as many of them involve behavioral manipulations. However, there are two specific ways to alter discounting through trial presentation and framing of questions. While the majority of DD studies present delays in terms of units of time (e.g., 6 weeks, 5 months, or 10 years), presenting the delay with a specific date instead can reduce the steepness of the discount curve [81, 82]. Conversely, labeling the delay in terms of days increases discounting [82]. It is possible that framing the delay as an exact date engenders more certainty of reward receipt than a more abstract time in the future or 1000 days. This effect may not be apparent in healthy populations, as ceiling effects may prevent decreases in discounting [83], or at shorter delays, given that as an event approaches, people tend to describe the time until an event in days and presenting this as a date makes it seem further away [17].

Another mechanism for reducing discounting is to highlight the mutual exclusivity of the immediate or delayed reward by stating that $0 will be awarded at the alternate timepoint [84]. For instance, in the explicit 0 paradigm, participants are asked to choose between $5 now and $0 a week from now ***or*** $0 now and $5 a week from now. Although this is technically always the correct scenario (if an immediate reward is chosen, there is nothing available later), explicitly reminding the participant that they will receive nothing later may increase attention to distal outcomes. This creates an "improving sequence" and enhances patience, thereby reducing discounting (for review, *see* [85]); although this hypothesis is refuted by the fact that the effect persists when participants are asked to consider past rewards, with more distal rewards being valued more highly [84]. An alternate hypothesis for this tendency is the *temporal attention hypothesis,* which suggests that the explicit zero addition shifts attention away from the present, making the future more relevant [80].

Other Considerations (Including Context Effects)

The breadth of studies in DD has provided incredibly useful insights into how presentation of monetary trials can affect responding in other ways beyond those described above. Dai et al. [86] identified reward contrast effects in DD wherein the value of the reward varies inversely with the surrounding context. For instance, if the standard option for one trial is $5000, followed by a standard option of $500, the $500 standard choice will be discounted more steeply than if it follows a standard option of $50. When the participant is presented with $5000 standard option first, it devalues the $500 following standard option, making it less desirable in the next trial. A previous investigation of DD parameters observed differences in discounting between presentation of ascending or descending fixed immediate rewards [87], which may also reflect contextual influence in responding. It is important to remember that the order in which monetary amounts and delays are presented may engender development of "reference points" which influence future decisions. These studies suggest randomization of trial presentation is essential and large variabilities in standard amounts may impact discounting.

In addition to the context of trial presentation, the context of the physical surroundings and state of the participant can influence responding. Often when patients are in therapy for impulse control disorders, they profess an intention to abstain from impulsive behaviors, whether it be gambling or overeating or substance use; however, when confronted with cues outside of the therapy setting, they may be unable to resist. Dixon et al. [88] found that pathological gamblers discounted monetary rewards more steeply when the task was administered in an off-track betting establishment than in the laboratory. This finding has two primary implications. First, it is important for researchers to appreciate that results observed in a therapy session, for example, may not completely represent actual behavior in the real world. Second, if testing a potential intervention in a disordered population, it may be advantageous to administer the task within a setting reminiscent of what the participant encounters in day-to-day life. Another important aspect of administration of the task in participants with maladaptive behavioral patterns is deprivation. Heroin-addicted individuals will discount rewards more steeply during mild opioid withdrawal [89], so the physical state of the participant should always be evaluated prior to performing the task for the most accurate results. In addition to the participant's physical state, there is also evidence that mood, stress, and exposure to affective stimuli can affect responding as well [90].

Beyond attention to trial presentation order and monetary reward amounts, how the amounts themselves are worded may also have effects. Any type of presentation which requires more cognitive load may generate a steeper discounting curve. Rewards that are concrete amounts, such as $10, are not discounted as steeply as "fuzzy" amounts, such as a handful of quarters [91]. The authors suggest that the cognitive load required to translate a handful of quarters into a concrete dollar amount decreases its value. Similarly, when reward amounts are presented as decimals, such as $9.50, they are also discounted more steeply than whole numbers like $10 [92]. Again, this tendency is thought to result from the extra processing necessary for the fraction presented as $9.50. This idea is also in keeping with the alterations in discounting observed by using a specific date for the delayed reward, rather than the less explicit number of days, weeks, or months described above.

Finally, several other modifications in trial presentation have effects on discounting behavior. These include adding delays to both choice options ($10 a month from now or $50 six months from now), making the delayed option "in addition" to what is available immediately ($10 now or $10 + $5 a week from now), or adding a third option ($5 now, $10 a week from now, or $50 six months from now), all of which have been shown to decrease discounting (for review, *see* [85]). Another method for reducing discounting is to instill a sense of trustworthiness of the investigator, which leads participants to believe there is a greater likelihood that delayed rewards will be received [93].

5 Conclusions

Delay discounting provides a well-tested method of obtaining behavioral data on impulsivity in both children and adults. One of the most beneficial aspects of the task is its malleability. Although the focus of this chapter is task use with monetary rewards in humans, numerous studies use other commodities, such as food or drug, which also generate hyperbolic discounting curves. The DD task is useful in preclinical studies, as well, allowing for great back translation, pharmacological manipulations, and genetic studies. However, it is this same malleability which can engender changes in results, either orchestrated or incidental. While this chapter provides an overview of the most researched task parameters, investigators should be aware of possible procedural influences when deviating from the more standard protocols detailed above.

Acknowledgments

I would like to thank Drs. Nicholas Grahame and Brandon Oberlin for guidance on DD task parameters and statistical analyses.

References

1. Mazur JE (1987) An adjusting procedure for studying delayed reinforcement. Quant Anal Behav 5:55–73
2. Alessi S, Petry NM (2003) Pathological gambling severity is associated with impulsivity in a delay discounting procedure. Behav Process 64(3):345–354

3. Johnson MW et al (2020) Sexual discounting: a systematic review of discounting processes and sexual behavior. Exp Clin Psychopharmacol. https://doi.org/10.1037/pha0000402
4. Miranda-Olivos R et al (2021) The neural correlates of delay discounting in obesity and binge eating disorder. J Behav Addict 10(3):498–507
5. Fields SA, Ramos A, Reynolds BA (2015) Delay discounting and health risk behaviors: the potential role of stress. Curr Opin Psychol 5:101–105
6. Wismans A et al (2021) The role of impulsivity and delay discounting in student compliance with COVID-19 protective measures. Personal Individ Differ 179:110925
7. Petry NM (2001) Delay discounting of money and alcohol in actively using alcoholics, currently abstinent alcoholics, and controls. Psychopharmacology 154(3):243–250
8. Bickel WK, Odum AL, Madden GJ (1999) Impulsivity and cigarette smoking: delay discounting in current, never, and ex-smokers. Psychopharmacology 146(4):447–454
9. Turner JK et al (2021) The phenotype of recovery V: does delay discounting predict the perceived risk of relapse among individuals in recovery from alcohol and drug use disorders. Alcohol Clin Exp Res 45(5):1100–1108
10. Sheffer CE et al (2014) Delay discounting rates: a strong prognostic indicator of smoking relapse. Addict Behav 39(11):1682–1689
11. Hoffman WF et al (2006) Neuropsychological function and delay discounting in methamphetamine-dependent individuals. Psychopharmacology 188(2):162–170
12. Cox DJ et al (2020) Delay and probability discounting in cocaine use disorder: comprehensive examination of money, cocaine, and health outcomes using gains and losses at multiple magnitudes. Exp Clin Psychopharmacol 28(6):724
13. Kirby KN, Petry NM (2004) Heroin and cocaine abusers have higher discount rates for delayed rewards than alcoholics or non-drug-using controls. Addiction 99(4):461–471
14. Kirby KN (2009) One-year temporal stability of delay-discount rates. Psychon Bull Rev 16(3):457–462
15. Anokhin AP, Golosheykin S, Mulligan RC (2015) Long-term test–retest reliability of delayed reward discounting in adolescents. Behav Process 111:55–59
16. Anokhin AP et al (2011) Heritability of delay discounting in adolescence: a longitudinal twin study. Behav Genet 41(2):175–183
17. Lempert KM, Phelps EA (2016) The malleability of intertemporal choice. Trends Cogn Sci 20(1):64–74
18. Bailey AJ, Romeu RJ, Finn PR (2021) The problems with delay discounting: a critical review of current practices and clinical applications. Psychol Med 51:1–8
19. Johnson MW, Bickel WK (2008) An algorithm for identifying nonsystematic delay-discounting data. Exp Clin Psychopharmacol 16(3):264
20. Myerson J, Green L, Warusawitharana M (2001) Area under the curve as a measure of discounting. J Exp Anal Behav 76(2):235–243
21. Epstein LH et al (2003) Comparison between two measures of delay discounting in smokers. Exp Clin Psychopharmacol 11(2):131
22. Smith CL, Hantula DA (2008) Methodological considerations in the study of delay discounting in intertemporal choice: a comparison of tasks and modes. Behav Res Methods 40(4):940–953
23. DeHart WB et al (2018) The effects of outcome unit framing on delay discounting. J Exp Anal Behav 110(3):412–429
24. Kirby K, Petry N, Bickel W (1999) Monetary choice questionnaire. Retrieved from PsycTESTS. https://www.apa.org/pubs/databases/psyctests/
25. Mazur JE (1997) Choice, delay, probability, and conditioned reinforcement. Anim Learn Behav 25(2):131–147
26. Xu S et al (2018) Differential effects of real versus hypothetical monetary reward magnitude on risk-taking behavior and brain activity. Sci Rep 8(1):1–9
27. Isen AM, Patrick R (1983) The effect of positive feelings on risk taking: when the chips are down. Organ Behav Hum Perform 31(2):194–202
28. Hertwig R, Ortmann A (2001) Experimental practices in economics: a methodological challenge for psychologists? Behav Brain Sci 24(3): 383–403
29. Camerer C, Mobbs D (2017) Differences in behavior and brain activity during hypothetical and real choices. Trends Cogn Sci 21(1):46–56
30. Johnson MW, Bickel WK (2002) Within-subject comparison of real and hypothetical money rewards in delay discounting. J Exp Anal Behav 77(2):129–146
31. Madden GJ et al (2003) Delay discounting of real and hypothetical rewards. Exp Clin Psychopharmacol 11(2):139
32. Madden GJ et al (2004) Delay discounting of potentially real and hypothetical rewards: II. Between-and within-subject comparisons. Exp Clin Psychopharmacol 12(4):251

33. Locey ML, Jones BA, Rachlin H (2011) Real and hypothetical rewards. Judgm Decis Mak 6(6):552
34. Bickel WK et al (2009) Congruence of BOLD response across intertemporal choice conditions: fictive and real money gains and losses. J Neurosci 29(27):8839–8846
35. Kang MJ et al (2011) Hypothetical and real choice differentially activate common valuation areas. J Neurosci 31(2):461–468
36. Whiteside SP, Lynam DR (2001) The five factor model and impulsivity: using a structural model of personality to understand impulsivity. Personal Individ Differ 30(4):669–689
37. Cyders MA et al (2007) Integration of impulsivity and positive mood to predict risky behavior: development and validation of a measure of positive urgency. Psychol Assess 19(1):107
38. Weafer J, Baggott MJ, de Wit H (2013) Test–retest reliability of behavioral measures of impulsive choice, impulsive action, and inattention. Exp Clin Psychopharmacol 21(6):475
39. Reynolds B, Richards JB, de Wit H (2006) Acute-alcohol effects on the experiential discounting task (EDT) and a question-based measure of delay discounting. Pharmacol Biochem Behav 83(2):194–202
40. Ortner CN, MacDonald TK, Olmstead MC (2003) Alcohol intoxication reduces impulsivity in the delay-discounting paradigm. Alcohol 38(2):151–156
41. Weafer J, Fillmore MT (2016) Low-dose alcohol effects on measures of inhibitory control, delay discounting, and risk-taking. Curr Addict Rep 3(1):75–84
42. Du W, Green L, Myerson J (2002) Cross-cultural comparisons of discounting delayed and probabilistic rewards. Psychol Rec 52(4):479–492
43. Mitchell SH (1999) Measures of impulsivity in cigarette smokers and non-smokers. Psychopharmacology 146(4):455–464
44. Rodzon K, Berry MS, Odum AL (2011) Within-subject comparison of degree of delay discounting using titrating and fixed sequence procedures. Behav Process 86(1):164–167
45. Mazur JE (1988) Estimation of indifference points with an adjusting-delay procedure. J Exp Anal Behav 49(1):37–47
46. Reynolds B, Schiffbauer R (2004) Measuring state changes in human delay discounting: an experiential discounting task. Behav Process 67(3):343–356
47. Horan WP, Johnson MW, Green MF (2017) Altered experiential, but not hypothetical, delay discounting in schizophrenia. J Abnorm Psychol 126(3):301
48. Jimura K et al (2011) Domain independence and stability in young and older adults' discounting of delayed rewards. Behav Process 87(3):253–259
49. Sutcliffe KR et al (2019) Relationships among academic procrastination, psychological flexibility, and delay discounting. Behav Anal Res Pract 19:315–326
50. Macaskill AC, Hunt MJ, Milfont TL (2019) On the associations between delay discounting and temporal thinking. Personal Individ Differ 141:166–172
51. Xu P, González-Vallejo C, Vincent BT (2020) Waiting in intertemporal choice tasks affects discounting and subjective time perception. J Exp Psychol Gen 149:2289–2313
52. Stam CH, van der Veen FM, Franken IHA (2022) Individual differences in time estimation are associated with delay discounting and alcohol use. Curr Psychol 41(6):3806–3815
53. Witt KM, Macaskill AC (2022) Choices about whether to wait: changes in delay discounting without changes in time perception. Behav Process 200:104696
54. Johnson MW (2012) An efficient operant choice procedure for assessing delay discounting in humans: initial validation in cocaine-dependent and control individuals. Exp Clin Psychopharmacol 20(3):191
55. Koffarnus MN, Bickel WK (2014) A 5-trial adjusting delay discounting task: accurate discount rates in less than one minute. Exp Clin Psychopharmacol 22(3):222
56. Rachlin H, Raineri A, Cross D (1991) Subjective probability and delay. J Exp Anal Behav 55(2):233–244
57. Johnson PS, Herrmann ES, Johnson MW (2015) Opportunity costs of reward delays and the discounting of hypothetical money and cigarettes. J Exp Anal Behav 103(1):87–107
58. Chapman GB (1996) Temporal discounting and utility for health and money. J Exp Psychol Learn Mem Cogn 22(3):771
59. Harless DW (1992) Actions versus prospects: the effect of problem representation on regret. Am Econ Rev 82(3):634–649
60. Green L, Myerson J, McFadden E (1997) Rate of temporal discounting decreases with amount of reward. Mem Cogn 25(5):715–723
61. Grace RC, Sargisson RJ, White KG (2012) Evidence for a magnitude effect in temporal discounting with pigeons. J Exp Psychol Anim Behav Process 38(1):102
62. Oberlin BG et al (2015) Monetary discounting and ventral striatal dopamine receptor availability in nontreatment-seeking alcoholics and

social drinkers. Psychopharmacology 232(12): 2207–2216
63. Cross CP, Copping LT, Campbell A (2011) Sex differences in impulsivity: a meta-analysis. Psychol Bull 137(1):97
64. Hosseini-Kamkar N, Morton JB (2014) Sex differences in self-regulation: an evolutionary perspective. Front Neurosci 8:233
65. Eubig PA et al (2014) Sex differences in response to amphetamine in adult Long–Evans rats performing a delay-discounting task. Pharmacol Biochem Behav 118:1–9
66. Smethells JR et al (2016) Sex differences in the reduction of impulsive choice (delay discounting) for cocaine in rats with atomoxetine and progesterone. Psychopharmacology 233(15): 2999–3008
67. Rosch KS, Mostofsky SH, Nebel MB (2018) ADHD-related sex differences in fronto-subcortical intrinsic functional connectivity and associations with delay discounting. J Neurodev Disord 10(1):1–14
68. Ferguson E et al (2021) Sex differences in associations between delay discounting and expectancies for alcohol analgesia. Exp Clin Psychopharmacol. https://doi.org/10.1037/pha0000517
69. Shibata Y (2013) Sex differences in the effects of disinhibition, perceived peer drinking, and delay discounting on drinking among Japanese college students. Personal Individ Differ 55(7): 766–770
70. Göllner LM et al (2018) Delay of gratification, delay discounting and their associations with age, episodic future thinking, and future time perspective. Front Psychol 8:2304
71. Liu L-L et al (2016) Age differences in delay discounting in Chinese adults. Personal Individ Differ 90:205–209
72. Read D, Read NL (2004) Time discounting over the lifespan. Organ Behav Hum Decis Process 94(1):22–32
73. Burrow AL, Spreng RN (2016) Waiting with purpose: a reliable but small association between purpose in life and impulsivity. Personal Individ Differ 90:187–189
74. De Luca CR et al (2003) Normative data from the CANTAB. I: development of executive function over the lifespan. J Clin Exp Neuropsychol 25(2):242–254
75. Jachimowicz JM et al (2017) Community trust reduces myopic decisions of low-income individuals. Proc Natl Acad Sci 114(21): 5401–5406
76. Kawashima K (2006) The effects of inflation and interest rates on delay discounting in human behavior. Psychol Rec 56(4):551–568
77. Ostaszewski P, Green L, Myerson J (1998) Effects of inflation on the subjective value of delayed and probabilistic rewards. Psychon Bull Rev 5(2):324–333
78. Peters J, Büchel C (2010) Episodic future thinking reduces reward delay discounting through an enhancement of prefrontal-mediotemporal interactions. Neuron 66(1): 138–148
79. Lempert KM et al (2019) Can delay discounting deliver on the promise of RDoC? Psychol Med 49(2):190–199
80. Naudé GP et al (2018) Temporal framing and the hidden-zero effect: rate-dependent outcomes on delay discounting. J Exp Anal Behav 109(3):506–519
81. Read D et al (2005) Four score and seven years from now: the date/delay effect in temporal discounting. Manag Sci 51(9):1326–1335
82. DeHart WB, Odum AL (2015) The effects of the framing of time on delay discounting. J Exp Anal Behav 103(1):10–21
83. Klapproth F (2012) The date-delay framing effect in temporal discounting depends on substance abuse. Behav Process 90(3): 420–423
84. Radu PT et al (2011) A mechanism for reducing delay discounting by altering temporal attention. J Exp Anal Behav 96(3):363–385
85. Rung JM, Madden GJ (2018) Experimental reductions of delay discounting and impulsive choice: a systematic review and meta-analysis. J Exp Psychol Gen 147(9):1349
86. Dai Z, Grace RC, Kemp S (2009) Reward contrast in delay and probability discounting. Learn Behav 37(3):281–288
87. Robles E, Vargas PA (2008) Parameters of delay discounting assessment: number of trials, effort, and sequential effects. Behav Process 78(2):285–290
88. Dixon MR, Jacobs EA, Sanders S (2006) Contextual control of delay discounting by pathological gamblers. J Appl Behav Anal 39(4): 413–422
89. Giordano LA et al (2002) Mild opioid deprivation increases the degree that opioid-dependent outpatients discount delayed heroin and money. Psychopharmacology 163(2): 174–182
90. Phelps EA, Lempert KM, Sokol-Hessner P (2014) Emotion and decision making: multiple modulatory neural circuits. Annu Rev Neurosci 37(1):263–287
91. Harman M, Kodak T, McKerchar T (2020) Effects of reward magnitude frames on measures of delay discounting in a hypothetical

money scenario. J Exp Anal Behav 114(2): 193–202

92. Fassbender C et al (2014) The decimal effect: behavioral and neural bases for a novel influence on intertemporal choice in healthy individuals and in ADHD. J Cogn Neurosci 26(11):2455–2468

93. Michaelson L et al (2013) Delaying gratification depends on social trust. Front Psychol 4: 355

Chapter 4

Lickometry to Circuitry: How Compulsion-Like Alcohol Drinking Microstructure Helped Discover Functional Differences in Salience Network Regions

Thatiane De Oliveira Sergio, Phillip A. Starski, and Frederic W. Hopf

Abstract

Problem alcohol drinking remains an enormous personal, social, and clinical challenge, and compulsion-like drinking (intake despite known negative consequences) can strongly contribute. Here we describe the lickometry method, a simple procedure (determining lick times) but, by analyzing many aspects of licking (or some other behavior), but with potential to provide important insights into underlying psychological and action strategies. We compare compulsion-like drinking with moderate challenge or higher challenge (ModChD or HiChD, quinine-alcohol) versus alcohol-only drinking (AOD). Classic studies suggest two main strategies during oral intake, more bouts of licking (indicating higher motivation), or longer (often faster) bouts (indicating higher palatability). In strong contrast, ModChD shows decreased variability of many lick measures, from lick speed to tongue shape, consistent with importance of automaticity in addiction. Further, and surprisingly, HiChD retains decreased tongue variability (and earlier bout start) seen with ModChD-versus-AOD, along with greatly disrupted intake. We relate these changes to regions of the Salience Network, which identifies and responds to important events. HiChD collapses mirrors inhibiting dorsal-medial prefrontal cortex (dMPF), while anterior insula cortex (AIC/ventral-frontal cortex) is linked to main value-driven action plan and also oral consumption. While AIC-dMPF likely work together, we find a novel dissociation in AIC-versus-dMPF function: AIC provides overall drive and firm commitment regardless of challenge level, while dMPF organizes action across time, but the battle for acting is won or lost in the dMPF.

Key words Alcohol compulsion, Lickometry, Anterior insula, Quinine

1 Introduction

1.1 Lickometry

Oral intake is a critical and essential part of life for most creatures, and neuroscientists have long worked to discover the multiple likely mechanisms that promote consumption. Indeed, over more than a half century, behavioral scientists have amassed many studies that examined changes in the so called "behavioral microstructure" of rodent licking in relation to conditions of ostensibly higher or lower motivation (e.g., higher or lower level of sugar, lower or

Melissa A. Cyders (ed.), *Translational Research Methods for Alcohol Use Disorders*, Neuromethods, vol. 201,
https://doi.org/10.1007/978-1-0716-3267-3_4,

higher bitter compounds in the sugar, *see* Subheading 2.2). Indeed, rodents and humans tend to consume in rhythmic patterns, and one can measure several aspects of responding, such as the speed of licking or chewing, or whether intake continues for longer or shorter time (called "bouts"). With several other excellent recent lickometry reviews (e.g., [1]), the intent of this commentary is to succinctly review the classic lickometry literature, then describe how our lab came to lickometry. Briefly, we sought to understand the critical, underlying psychological states that drive alcohol intake, and to discover the brain systems mediating these pathological urges. As detailed below, lickometry patterns can, with care, be used to infer some aspects of underlying intake motives. Indeed, we found several unexpected changes in response patterns related to alcohol drinking, which ultimately took us in an important new direction: the idea that aversion-resistant responding involves choosing to act with more automaticity and stereotyped action. Importantly, we also speculate that keeping attention on the main goal is an important part of ignoring the negative consequences (*see* Subheading 3).

1.2 Compulsion-like Alcohol Drinking in Humans and Rodents, Including Possible Common Brain Circuit

Before diving into lickometry analysis, we first return to better address the scale and scope of human problem drinking. In particular, it would be helpful to consider the strengths and limitations of rodent compulsion-like drinking models, used along with alcohol-only regular intake, in order to care about rat lickometry findings. More generally, Alcohol Use Disorder (AUD) extracts ~$250 billion/year in the USA, with many medical and social harms [2–9]. Also, problem drinking in women has risen dramatically in recent years [10–12], and women can have greater alcohol problems [13–15]. Indeed, compulsion-like alcohol drinking (CLAD), where intake persists despite negative consequences, can be a major obstacle to treatment and a strong driver of excessive intake [16–25]. However, even with significant harms, there are few AUD pharmacological treatments [26, 27], and a critical unmet need to develop better, personalized therapies [28, 29].

To improve our understanding and ability to treat compulsive drives, in particular to understand critical brain mechanisms of CLAD and AOD, we used rodent models. Rodent models provide many advantages, including precise modeling of different aspects of behavior. In addition, rodent work also is essential for precise, projection-specific mechanistic intervention to assess functional importance of specific projections for different behaviors. Also, while human studies provide crucial information about behaviorally relevant brain activation related to AUD, rodent in vivo recordings are essential to capture actual firing patterns across different neuronal populations, providing a more rigorous view of what is encoded (relative to the slow and uniform human fMRI). Another challenge for human studies is disentangling biological versus social influences [30–32], making rodent studies essential to understand more basic underlying factors.

Importantly, our and other's rodent studies suggest that a similar brain circuit mediates compulsion-like responding for alcohol in humans and rodents through regions of what is termed the Salience Network (SN), which mediates rapid identification and responding to important situations [33–37]. The SN is likely directed by the anterior insula cortex (AIC) and dorsal anterior cingulate cortex, a part of the medial prefrontal cortex (dMPF). (This is likely oversimplified, in part because MPF [38] and AIC [33, 39] in humans have multiple subregions; thus, for some human studies, we use "MPF.") Overall, human AIC is linked to many aspects of alcohol drinking, and addictions and emotion regulation (adaptive and disordered) more generally [33], along with dMPF [40, 41].

Interestingly, we find in rats that specific AIC projections (e.g., AIC-ventral striatum and AIC-Locus Coeruleus area) promote CLAD in male rodents, with little impact on AOD [42, 43], and AIC mediates CLAD but not AOD in mice [44]. Importantly, compulsion-like responding for alcohol involves a similar AIC-dMPF-striatal circuit in women and men with AUD [45, 46] (*see* Subheading 3). Indeed, human AIC activity is linked to many aspects of alcohol drinking [33], including where alcohol cue- [47, 48] and negative affect-related [49] AIC activation predict real-world drinking (see also [50]), and AIC activation to alcohol images predicts later transition to heavy drinking [51]. Also, AIC changes are linked to worse behavioral control in AUD [52, 53], while therapies that reduce drinking decrease AIC activity and/or connectivity (e.g., [54, 55]). Thus, while we have tried to operationalize compulsion-like responding as willingness to continue drinking despite anticipated aversive consequences, and we believe this ability to ignore/mitigate the impact of negative possibilities (real or anticipated) is fundamental to good emotional regulation (*see* Subheading 3). In addition, the similar AIC circuit for human and rodent for alcohol compulsion gives strong support that our rodent CLAD model has some relevance to humans (also observed for NMDA receptor regulation of drinking cross species, *see* [23, 56]). For other discussion of the relation between human compulsion for alcohol for rodent CLAD, *see* [23, 24, 57], and a comprehensive discussion in a separate review from De Oliveira Sergio et al. (submitted).

Other findings also suggest that AIC-MPF are important for overcoming cost to allow action, with activity related to discounting and overcoming of task difficulty, risk, and other costs [58–63], and promotion of risk seeking [63–65]. dMPF-NAcb regulates decreased sensitivity to loss, which promotes risk [66]. AIC-MPF circuits are also critical for action that occurs despite effort, delay, risk of loss [58–60], or urge to stop [67]. AIC-MPF are less active when dieters overcome temptation [68]. Thus, AIC-MPF are broadly implicated in overcoming challenge to allow action, of which compulsion-like alcohol directing is one type. One specific aspect of compulsion-like responding is the presumption that there is some aspect of cost-ignoring that is central for maintaining responding. In this light, MPF activity reflects directing attention away from pain to decrease subjective intensity [69] (*see* also [70]), and AIC-dMPF co-activate when resisting a belief change [71], trusting with possible betrayal [72], and imagining higher reward which discounts delay costs [73]. AIC-MPF are also important for sustained attention [61, 74], including continuing attention to reward under distraction [75] and sustained vigilance during anxiety [76] or uncertainty [77]. Together, these studies implicate AIC-MPF in managing the impact of negative information to maintain challenge-resistant action.

2 Methods: What Is Lickometry and How Can We Use It?

2.1 Technical Methods Used to Detect Licking

Lickometry is quite simple to perform, as small changes in electrical current detects a lick when the rat tongue touches the metal spout. For this commentary, we're going to take some specific examples from our published studies [78, 79] (in Figs. 1, 2, and 3 here) to demonstrate how lickometer measures can provide novel insights into underlying action strategies. For these studies, rats licked for alcohol while in a standard operant box (normally used for lever press studies), and a rat has to rear up slightly in order to insert the nose into a space to lick; tongue contact on the metal alcohol spout completes a circuit through the metal floor the rat stands on. While very simple technically, analysis and interpretation of lickometry is where the art is, and the following sections are designed to help better understand how different lick measures suggest different internal action strategies, ready to be activated when access for intake begins.

To make lickometry even simpler, we have more recently used a capacitance-based lickometer system (run from an Arduino, *see* biorvix: [80, 81]), and when the rat licks the metal spout, there is a very small change in electrical charge. This system has several advantages over the method published in [78, 79], including low cost, technical simplicity, and it does not require rodents to be adapted and trained in operant boxes, making experiments easier and quicker. Importantly, the electrical change with this system is small enough that it does not disrupt the ability to record single-unit (individual neuron) neuronal activity with microwires in vivo. Indeed, several groups have recorded firing during licking using very similar lickometers [82, 83], including in the AIC [84, 85]. In addition, this method does not require a metal floor (needed in operant box to detect closing of circuit when tongue touches the licker). However, the method in [78, 79] perhaps has a more complicated physical response requirement than the seemingly simpler, home-cage bottle licking. Nonetheless, our studies in [80, 81]

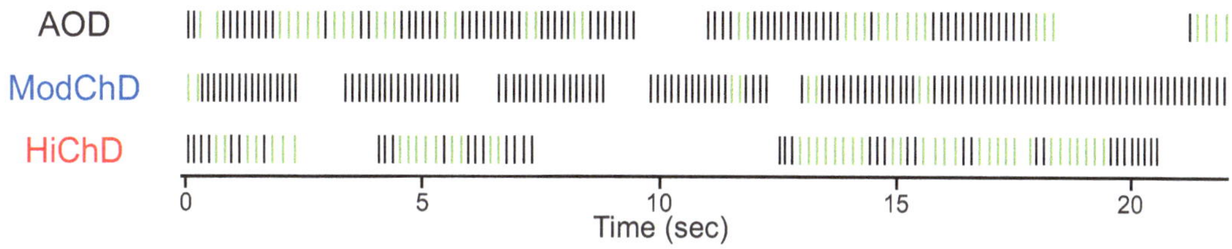

Fig. 1 Less variable lick speed with ModChD versus AOD. Examples of trains of licking for AOD (top), ModChD (middle), and HiChD (bottom). Black: lick with <175 msec inter-lick interval (ILI) on each side; green: lick with ILI longer than 175 msec on at least one side of the lick

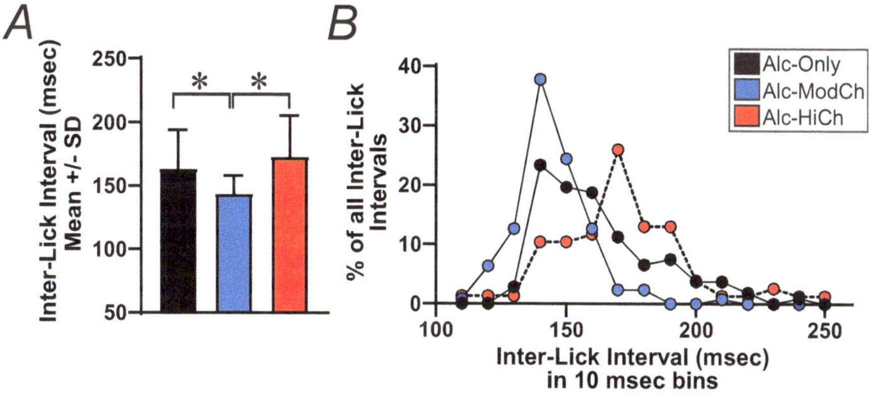

Fig. 2 Less variable lick speed with ModChD versus AOD. (**a**) average and standard deviation of inter-lick intervals of responding shown in Fig. 1. (**b**) Histogram of ILI distributions, more narrow with AOD vs others

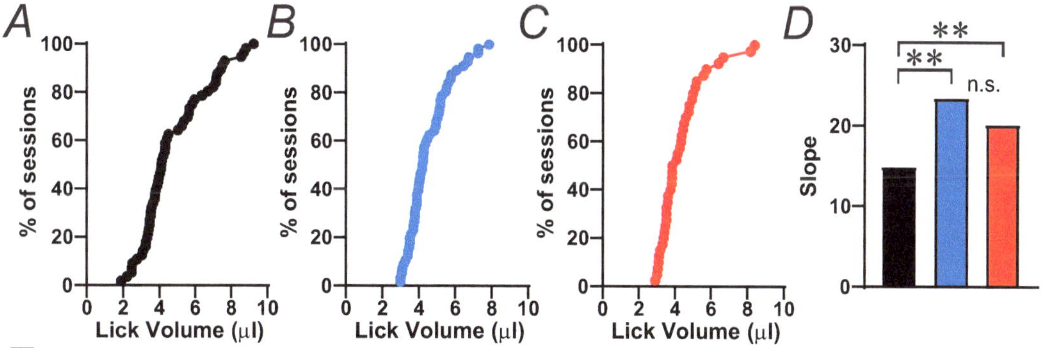

E

	Range	lowest session	10% of sessions	middle session	90% of sessions	highest session	90% minus 10%
AOD	7.3	1.9	2.7	4.1	7.4	9.2	4.7
ModChD	4.9	3.0	3.2	4.2	6.3	7.9	3.1
HiChD	5.5	2.9	3.1	3.8	5.7	8.4	2.6

Fig. 3 Lick volume is less variable across ModChD and HiChD versus AOD sessions

with home-cage lickometry find some consistent patterns with earlier work, including where HiChD reduced overall intake and licking speed, had greater variability in timing, and led to bout disorganization (*see* below). Also, with the large difference in body size between long-term drinking females (~450 g average) and males (~750 g average), the basic geometry and execution of licking may differ, but in fact we find limited (although important) differences in licking measures between females and males (biorvix: [80]).

Also, while we focus on studies of analyzing licks during free intake, it is important to note that lickometry methods have much more extensive use. One early sophisticated lickometer system, the Davis box, had a system with 8–12 lick spouts that were movable, and rats had very brief access to lick each. In addition to measuring responding for different substances in the same session, the method can be used to separate many aspects of taste processing; e.g., cocaine aversion decreases sugar preference (taken to indicate anhedonia), and a recent study [86] used this system to show that cocaine does not change the ability to discriminate the sensory properties of different tastes, but slows the latency to respond (a decrease in motivation). Other studies have utilized lickometry to measure tongue force and other more sophisticated indicators [1], showing the broad utility of this technique.

2.2 Basics of Classic Lickometry: Bout Number Versus Bout Length Strategies

From first principle, there are a number of different lick-related measures one could assess. Licking mostly happens at some average speed (partly determined by the biomechanics of licking [87]), which could change across time; one can also examine the variability of licking speed, e.g., having slower and faster moments of licking in the same lick train. In addition, one can examine how long responding is sustained, e.g., with longer or shorter trains of licking for alcohol (Bout Length, Blen). Figure 1 shows examples of licking for alcohol during the first ~20 s of a drinking session (with ~80% of intake in the first 5–6 min). We first can notice there are trains of licks which come in different lengths, with ~1 s between lick trains. In the middle example of Fig. 1, there are four shorter bouts, then a bout more than twice the length of the previous bouts. Another important measure is Total Intake, the amount of consumption an individual actually achieves, which is often the most important measure we are interested in, e.g., in our CLAD non-lickometer work [42, 43, 56, 81, 88]. For alcohol, this is the g/kg intake, which is ~0.8–1.2 g/kg/5 min for alcohol-only in our hands.

Importantly, classic (and more recent) studies have often converged on the idea that changes in Bout Number (Bnum), without changes in Blen, are more reflective of changes in motivation, while longer Blen indicates greater palatability [89–93]. Patwell et al. [90] summed it up well: "Analysis of licking microstructure can be used to dissect the effects of different reinforcer properties on consumption. For example, total number of drinking bouts and average bout length are frequently interpreted as proxies for incentive and palatability, respectively." Similarly, in one 2000 review [94], alcohol intake level in rats could be increased or decreased (with increasing costs, or in some genetically preferring lines), but almost all manipulations altered Bnum and not Blen. Indeed, only adding sugar to alcohol increased Blen, sometimes two- to three-fold. In another example [90], some rats lever-press for alcohol and drink a lot (Drinkers), while other rats press a lot but drink much less (Responders). Responders on average have only a single bout of alcohol intake (Bnum = one), compared to ~ four Bnum for drinkers, but Blen is similarly long in Responders and Drinkers. Perhaps in concurrence, having longer bouts is critical to successful higher-challenge alcohol drinking (with similar length and other measures across drinking conditions) [79]. In addition, we find that females have significantly longer alcohol bouts than males (biorvix: [80]). Finally, alcohol-preferring selected mouse lines can have different drinking patterns, with longer bouts in HDID1 and more bouts in HDID2 [95].

We are open-minded about this classic interpretation of Bnum versus Blen for motivation versus palatability. Certainly, these measures can represent the proposed underlying constructs under some conditions, but, as we will see below, it is important to measure and compare across as many lick parameters as possible, and to have caution in interpretation of underlying drives. Indeed, a recent study found that lickometry differences overall (from a machine learning model) predicted decreases in mouse alcohol intake with AIC-dorsolateral striatum [96]. Another important point is that the magnitude of change in a measure might not be large, but what is important is having indicators that may show a shift in internal action strategy.

2.3 Other Lickometry Measures: Uncovering Unpredicted Response Patterns when Drinking Alcohol in the Face of Challenge

With the likely importance of compulsion-like drives for alcohol in at least some human problem drinkers (*see* Subheading 1.2), we sought to determine whether lickometry and microstructural analyses could provide insight into possible shifts in underlying action strategy. For this, we compare alcohol-only drinking (AOD, ostensibly in the absence of challenge) with drinking alcohol in the face of negative/aversive challenge, including moderate challenge (ModChD, 10 mg/L quinine in alcohol, where intake level is not reduced) and higher challenge (HIChD, 100 mg/L quinine in alcohol). In particular, we use several examples of data analyses taken from our published broader, unbiased analyses across different licking behaviors [78, 79]. These example findings are meant to show the important, unexpected, and likely translationally relevant insights into different psychological states and action strategies that we believe are utilized during challenged drinking versus AOD.

2.3.1 Lickometry: Analysis of Licking Speed

Classic rat studies often relate changes in licking speed to shifts in drive for a consummant, including where faster responding along with longer bouts are taken to indicate higher palatability [78, 79, 89, 91, 92, 97]. We reasoned that, if ModChD reflected higher motivation for alcohol (willing to expend cognitive/emotional effort to overcome negative to get alcohol), we might see faster responding and perhaps longer bouts relative to AOD. Instead, we found that Blen and Bnum were not on average different, and, strikingly, found that ModChD licking was significantly less variable than AOD (in addition to being slightly but significantly faster).

To assess licking speed, we primarily use a measure called the Inter-Lick Interval (ILI), which is the amount of time between two successive licks. Technically, licking speed would be the rate of licking, e.g., 200 ms between licks is a rate of 5 licks/s. We prefer ILI, since it is widely used and still intuitive when figuring out changes in lick timing (300 ms between licks is slower than 100 ms between licks). In Fig. 1, we change the color of licks: licking faster than 175 msec between licks is shown in black marks; slower licking is switched to green (if a lick has a > 175 ms ILI on at least one side). Figure 1 gives a visual example of the less variable lick timing across ModChD intake compared with AOD. Figure 2a shows a bar graph with the average and standard deviation for all ILIs within bouts in Fig. 1, separated by drinking condition. ModChD licking is significantly less variable than AOD, and slightly but significantly faster (briefer ILI), as in our published studies [78, 79].

Also, because lickometry responses are often not normally distributed, it can be helpful to examine the distribution to better understand what the raw data looks like. For example, the AOD raster in Fig. 1 has 104 ILI data values, ranging from 128 to 365 ms. One can bin these ILI values and ask, e.g., how many ILIs are from 121 to 140 ms (19), how many ILIS are from 141 to 160 ms (43), etc. In Fig. 2b, the X axis shows ILI values, grouped into bins 10 ms wide. The Y axis shows the percent of all ILIs that have that particular bin-value of ILIs. This is done separately for each drinking condition. It is also possible to see that the distribution of ILIs across the X axis is broader across AOD and HiChD compared with ModChD (*see* Fig. 2b), another way to demonstrate the reduced variability in response timing relative to AOD (described more fully in [78, 79]). In addition, the HiChD data show part of the evidence indicating that response organization and timing is dramatically disrupted during drinking under higher challenge. Indeed, HiChD has a ~ 50% drop in intake, ~40% drop in bout length, and ~ 25% drop in bout number, when compared with AOD or ModChD (which have the same level of these measures) [78, 79, 98]. Also, more variable licking speed is seen with other challenged drinking conditions (water or sugar with quinine [92, 99], LiCl sickness [100]).

Also interesting is a recent study [101], where mice were exposed to repeated alcohol vapor, and, along with greater drinking, showed a greater variability in licking speed, even with higher alcohol intake (and longer BLen). Perhaps this could represent dMPF disruption by high alcohol exposure, so that bottom-up urges (e.g., from amygdala) drive intake, although the authors note that intoxication itself might contribute. Thus, studies could examine whether alcohol vapor leads to more variable responding during limited access where intoxication is not achieved.

More generally, studies examining front-loading infer that licking is faster: in a recent study, rats show strong front loading, with ~46% of intake in the first ~4 min of a 30-min session [102], similar to how our rats drink [78, 79] (biorvix: [80]). Importantly, female front-loading was ~80% larger than in males, while shortening the intake session (to 15 min) raised male front-loading to near female levels [102]. Thus, important aspects of motivation can be assessed without lickometry, although we consider that shifts in variability (next section) might be challenging to detect with only intake per min. For example, aspects of licking patterns (especially faster drinking) indicate higher alcohol intake in non-human primates [103] and greater drinking problems in humans [104].

2.3.2 Lickometry: Analysis of Lick Volume, a Measure of Tongue Control

Each drinking session has an average lick volume, usually determined as total volume intake divided by total licks and given as microliters per lick. Essentially, it is a measure of tongue control, including tongue shape, which seems esoteric, but turns out to be immensely valuable, as we will describe. In addition, Fig. 3 shows the use of another method, the cumulative distribution, as a useful visual tool that complements averages, histograms, and scatterplots [78, 79]. Thus, we can sort the Lvol values across the 56 AOD sessions in order from smallest Lvol (1.9) to largest Lvol (9.2) values, which become the ~3% and 100% of AOD sessions ranked by Lvol. This gives us a cumulative distribution for each drinking condition separately, and demonstrates visually that Lvol values under ModChD and HiChD are significantly less variable than AOD. In the table in Fig. 3, even just the Range (leftmost number column) of Lvol values across sessions is noticeably narrower for both compulsion-like drinking conditions relative to AOD.

Together, these different methods show that, even though the average Lvol value did not differ across drinking conditions [78, 79], it is possible to detect robust changes in Lvol variability. Also important is that the pattern of changes in Lvol across drinking groups is markedly different from that in Fig. 1, where licking speed was significantly faster and less variable for ModChD versus AOD, but HiChD shifted to slower and more variable licking versus ModChD. In addition, both HiChD and ModChD bouts start about 40 s earlier in the drinking session versus AOD (where

80% + intake occurs within ~6 min), a sign of a shift in overall plan in both ModChD and HiChD. Thus, even though HiChD has less and shorter bouts, it retains several behavioral patterns similar to ModChD, which we address more in Subheading 3.

2.4 Other Caveats to the Classic Model

There are other exceptions to predictions from the "classic" model. One group found that alcohol dependence in mice increased alcohol intake, with more intake, larger Bnum, and longer (Blen) and faster bouts, with no change in Lvol; this is perhaps similar to a combination of the two classic mechanisms in Subheading 2.2 [105] (also seen in [99]). Higher sucrose increases bout length but not speed [92]. Scopolamine in the dMPF reduces sugar intake, with more, shorter bouts, and no change in lick rate [106]. Also interesting is where ablation of a particular type of GABA neuron (parvalbumin+) in the dorsolateral striatum in mouse, which decreases AOD and CLAD ~35–40% [107]. AOD responding is faster, with fewer but longer bouts, and greater variability in Blen (but no changes in water or sugar intake), while for CLAD, responding is faster, with fewer bouts and no changes in Blen. Also, inhibition of orbitofrontal cortex [108] altered lick volume but not lick speed, although it did result in more, shorter bouts (with no change in overall intake). Finally, lick volume can decrease in relation to higher aversion [92, 99, 100], but also for pleasant tasting drinks ([91, 109], but *see* [97]). Thus, it has been more widely recognized that measuring multiple factors for a given behavior is important for generating a balanced and comprehensive understanding of any changes.

3 Results and Discussion: What Findings Can Be Inferred from Lickometry?

We use lickometry to attempt to uncover internal psychological mechanisms, and related action strategies, that are used during specific types of maladaptive alcohol drinking, especially compulsion-like drinking under different levels of aversive challenge [78, 79]. ModChD intake is not reduced relative to AOD, but the same quinine level in water is greatly avoided, suggesting aversion-resistant drive for alcohol. While we had predicted faster responding and longer bouts with ModCHD (indicating greater motivation and drive), but instead found that ModChD shows decreased variability in many response measures [78]. While unexpected, this agrees with the suggested importance of automaticity for habit and compulsion [25, 110, 111]. Accordingly, we proposed the Head Down and Push model [78], where decreased response variability reflects a psychological strategy with "internal" attention focused on stereotyped responding; this may minimize the need to attend to (and be impacted by) adverse consequences present during action. However, high enough challenge can reduce

CLAD intake level [24, 78, 79, 98], perhaps reflecting inability to ignore cost, which disrupts responding despite strong reward drive. Also, we looked for behavioral indicators that could reflect better or worse ability to maintain consumption despite stronger challenge [79]. Interestingly, even though HiChD strongly reduces intake, it retains several behavioral changes seen with ModChD, including less variable lick volume across sessions and earlier initiation of bouts within sessions. As detailed below, these suggest that rats maintain a greater global commitment to drinking under aversive consequences, separate from challenge level. Thus, AIC and dMPF likely work together under many conditions [33, 37, 78, 79, 112, 113], including for alcohol [42, 43, 114–116]; our studies provide important new insights into the different roles of AIC and DMP as they generate motivated action.

3.1 Lickometry Suggests Different AIC Versus dMPF Contributions to Alcohol Intake

We found the interesting observation that HiChD drinking retains better tongue control and sooner responding, which are seen with ModChD, but otherwise has disrupted organization and timing. Importantly, a relatively large literature supports the idea that AIC (and nearby ventral-frontal cortex) has been linked with encoding the main value-driven action plan, and for oral intake (including tongue shape) (which makes sense since consumption is likely a persisting, high-interest activity for all living things. In contrast, dMPF has been linked to moment-to-moment action organization, including lick timing and maintaining bouts of responding. Indeed, this literature supports a double dissociation, with AIC for main-plan/tongue control, but not action organization across time, with dMPF for organized responding but not for tongue control. This includes measures of tongue control, which have been associated with ventral-frontal cortex [108, 117–122], but not dMPF [106, 118, 119, 122, 123]. In contrast, disrupted responding under higher challenge was similar to more variable action timing and action fragmentation with MPF inhibition [106, 108, 120, 123–128], but not ventral frontal cortex (Stuss, 2003 #969; Stuss, 2005 #968; tested in AIC in [127, 129], and *see* [108] in Suppl.Disc). Importantly, there are also corroborating studies of medial versus ventral prefrontal differences in humans [130, 131]. Also, more generally, dMPF and ventral frontal cortex play different roles in decision-making, with dMPF for physical effort, with OFC regions mediating reward-action value and also time delays [132–136].

One major reason these findings are interesting to us is because our results suggest a novel hypothesis and dissociation in AIC and dMPF function within Salience Network promotion of alcohol intake. In particular, as shown in Fig. 4, HiChD have less variable lick volume and sooner responding, similar to ModChD, suggesting that AIC maintains its drive and commitment to responding even in the face of higher challenge. In contrast, the disrupted responding under HiChD, with greater variability in timing and

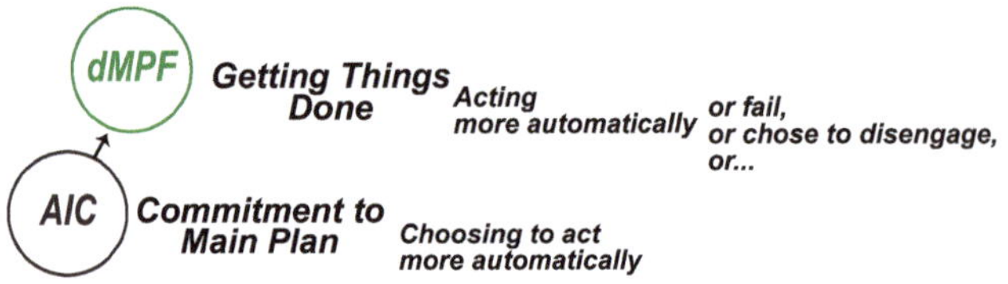

Fig. 4 Model of AIC versus DMP: AIC for main plan (including oral consumption); DMP for nuts and bolts of getting things done

shorter bouts, is quite reminiscent of what happens when dMPF is inhibited. It has long been appreciated that AIC and dMPF often activate and likely work together, with it commonly described that the AIC is the input and integrator of the Salience Network, and dMPF is the motor output side; across many studies, AIC is associated more with more overall considerations, with dMPF for the nuts and bolts of getting things down (*see* Fig. 4). However, the exact dissociable contributions of AIC and dMPF have not been clear. Here, our studies suggest that AIC provides overall drive regardless of challenge level, and dMPF organizes action across time. Importantly, even when AIC maintains firm commitment, the battle for acting (or not) is won or lost in the dMPF (less variable under ModChD, greatly disrupted under HiChD, perhaps due to inability to maintain coherent mPFC activity. In this vein, DMP function is impaired by stress [137, 138].

These findings concur with the central role of AIC, MPF, and SN for focusing attention on most important features of the environment and resulting motivated action, with less role during more certain, simple, and routine actions [34, 39, 61, 139]. Nonetheless, human studies also concur with dissociating AIC and MPF more for could be called more "overall" versus "proximal" aspects of behavior (mirroring AIC closer to main plan, and DMP closer to actual nuts and bolts of carrying out action, *see* Fig. 4). At a more basic level, AIC signals both aversion and reward, while MPF encodes only aversion [140]. Also, AIC reflects salience for both effort and reward, while MPF signals effort prediction errors and cost avoidance [141]. Other findings relate MPF more specifically to action, e.g., where AIC is important for both action- and cognition-related behavioral control, with MPF only for action-related control [142], and where MPF better predicts action choice and reaction time than ventral frontal cortex [143]. In addition, while AIC and MPF are both activated during regret, AIC reflects negative affect, with MPF-nucleus accumbens (NAc) encoding this and also the desire to change choice [144]. Furthermore, during unexpected emotional events, AIC provides a sustained signal of behavioral salience of events, while dMPF tracks attentional switching in relation to such events [112]. In addition, AIC encodes outliers while MPF signals variability [145], suggesting an AIC focus on the most important events, with MPF encoding all features within an environment (*see* [38, 146]). While the relative contributions of AIC and MPF likely depend on context, there is substantial evidence that AIC can reflect more general and MPF more proximal drive to evaluate and approach salient rewards (*see* [39]).

In addition, when considering possible AIC activity contributions and its encoding the "main plan," while AIC and SN are critical for initiating many motivated behaviors (*see* [33]), there is good evidence that the AIC plays a central role in sustaining motivated responding. Persistent AIC signals reflect the main goals of a task [74, 147] and persisting emotion [112] in humans, and the long-term reward value of a context in non-human primates [148]. Although there are quite limited studies in rodents, Guillem and colleagues [149] show hours-long plateau firing across a 3-h cocaine intake session in half of AIC cells, and AIC maintains cue responding across longer delays [150]. Also, the long-term reward study [148] found that AIC was the only region showing strong context-long value signals, and Pribut and colleagues [151] note, for cocaine in rats, that AIC is the main region among the many they have examined with more long-term firing changes (not directedly related to movement). Thus, AIC can support both initiation and sustaining of motivated responding. To identify AIC activity patterns that promote intake, we examined AIC in vivo firing during AOD and CLAD in male Wistar rats (biorvix: [81]). With data from 15 rats, many AIC neurons showed sustained, session-long changes in firing, whether increased or decreased. Importantly, only cells with stronger firing at the onset of licking had higher, plateau, firing for ModChD or HiChD versus AOD. The importance of initial firing, then sustained responding, comports with our findings [78, 79] that rats quickly evaluate the licking condition (AOD or challenge-resistant) and rapidly adjust their action strategy for licking (e.g., as demonstrated by rats starting bouts significantly sooner under challenged drinking vs. AOD).

However, it is likely that DMP, at a minimum, plays some important role in sustained, importance-direct action. AIC-MPF are critical for task set maintenance in humans [61, 147], allowing consistent reward-directed action across a session. MPF is also considered to initiate temporally extended and sustained action [143, 152], with stronger MPF activity likely required for persistent reward seeking under challenge [62]. AIC-MPF also contribute to sustained attention [74, 76] and tonic alertness under uncertainty [77]. Both AIC [153] and DMP cells [129, 154] fire for licking. Importantly, in rats, DMP cells continue to fire during licking even when reward delivery stops briefly [155], and gustatory insula food-related activity continues when sated [156]. Thus, AIC-MPF are well poised to support longer-term, protracted strategies for action somewhat detached from moment-to-moment events.

3.2 AIC May Have Two Pathways That Separately Regulate AOD Vs. CLAD

In addition to predictions about AIC-dMPF interactions, from lickometry and other studies, we also have a second model (*see* Fig. 5) suggesting that the AIC has two parallel pathways, one for CLAD and one for AOD. We find that AIC projections to subcortical areas (nucleus accumbens, NAcb, Locus Coeruleus Area) potently regulate CLAD, while these projections have no impact on AOD or sweet fluid intake [42, 43]. In strong contrast, overall inhibition of AIC (with locally-infused) GABA receptor agonists greatly reduces both AOD and CLAD (like intake of other intoxicants [157–159] and highly palatable food [160]). Along with other findings, and as shown in Fig. 5c, d, we hypothesize that AIC-subcortex selectively regulates challenge- and arousal-related behaviors, including CLAD. We then speculate that AIC projections to cortical areas regulate AOD; in particular, we are testing the importance of AIC-to-posterior insula (PIC), since inhibiting rat pINS decreases alcohol drinking [161], and PIC **projections** are quite large [33, 162], and AUD is associated with greater resting AIC-PIC connectivity in women [45] and men [163].

While this model remains highly speculative, the functional importance of specific AIC projections has received some study in rodents, with suggestive human work affirming some basic findings (*see* [33]). Interestingly, a recent study from humans converges upon the idea that AIC may have multiple mechanisms at the same time [164], and especially where interoception (sensing the body, a fundamental function of AIC) is more fundamental, and where emotional regulation involves AIC connectivity with striatum (AIC-subCX) (*see* Fig. 5e). Indeed, AIC connectivity with Nacb has been linked to compulsion for alcohol [46], and also finding pleasure in violence or some criminal acts [165, 166]. We strongly speculate that one core function of the Salience Network is

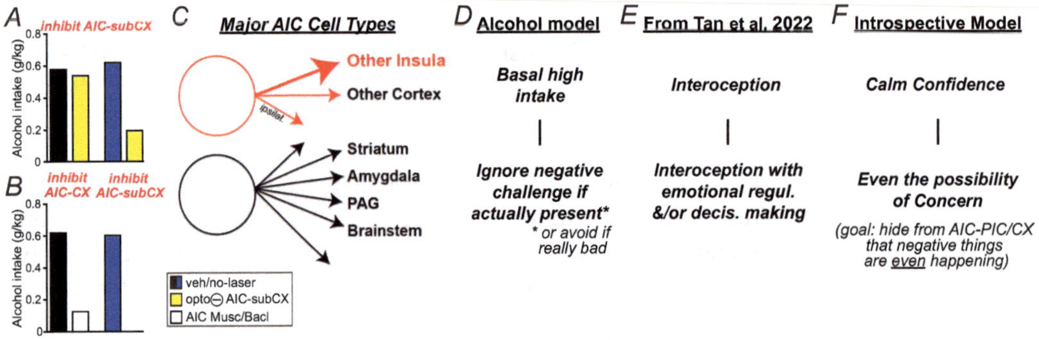

Fig. 5 Model of two AIC pathways for AOD versus challenged drinking, and confidence versus concern more generally; described in main text. Not shown in AIC-to-thalamus

to allow action to persist under challenge or stress, especially to prevent stress from rising high enough that cognitive function is disrupted. Thus, good emotional regulation (dampening arousal), especially through AIC, is essential for maintaining adaptive responding under challenge; this could be for heroic purposes (running into the burning building to save the baby) or dastardly (driving drunk even when you hit someone the week before), and many studies implicate AIC in the expression of discounting, ignoring, and a variety of other "emotional regulation" activities in humans [33].

Thus, Fig. 5 lays out several parallel conceptions of the function of a more basal vs. challenge-addressing AIC circuits. The more basal (which we propose is carried by AIC-PIC and other cortical areas) would promote AOD and other positive outcomes, while the AIC-subCX is recruited when there is actual challenge to deal with. One interesting note is that AIC inhibition does not impair sucrose intake in many studies [42, 43, 114, 167], which is unexpected because it seems like a high-motivation condition. We believe this reflects where alcohol is a high reward, and, importantly, the brain considers it worth fighting for (should the need arise), while sucrose is more basic. One possibility for AIC regulation of AOD is to make sure that, should a challenge appear, the AIC-PIC is within the circuit to quickly discount and ignore.

A pre-emptive readiness to ignore is seen in other habit-related areas, such as ventromedial PFC (VMP). For example, in a T-maze task that leads to habit learning with repeated training, the ventromedial PFC shows a plateau firing pattern that is sustained across the trial [168]. Importantly, after pairing the reward with sickness (LiCl), both the ability to keep responding (to ignore the conditioned sickness) and the VMP plateau firing decrease across days. Importantly, this plateau firing occurs before any challenged sessions, and we interpret this as a pre-loaded strategy, being ready to ignore any negative challenge (should it occur).

Finally, it might seem a disjunct that AIC encodes the main reward-related plan, and is also a primary area for oral action. Clearly, consumption is broadly of importance to living things. It has also been noted how important it is to have fast oral evaluation, to quickly assess taste and spit out if needed, or start licking very quickly [84, 85, 169], with taste sensed within one lick cycle [170]. On the other hand, it is also important to note that inhibiting AIC or AIC projections does not impact sweet fluid [42, 43, 114, 167] or water [160, 171] intake, simpler operant responding [64, 157, 172], or locomotion [43, 114, 160, 171, 173, 174], suggesting important selectivity in AIC contributions to motivated responding.

3.3 Conclusion

In conclusion, we have described studies from our work and from others which used lickometry, a technically simple technique long used to try to infer underlying motivation from differences in response patterns. Using work from our lab and others, we provide examples of the valuable types of insights that could be gained from lickometry, as long as caution is used in interpretation. In particular, we describe how these and related studies have given us what we think are important insights into how the AIC and DMP may play more specific roles in carrying compulsion-like consumption of alcohol, and many other motivated behaviors. Finally, we use these data to provide additional thoughts about the "purpose" of AIC contributions to drinking, especially where it mediates behaviors where one will expend cost (should it arise) to get a particular reward, and how AIC may have different pathways, one for more basic positive experiences versus, and the other for emotional regulation when challenge is present to allow action (to prevent high arousal which could impair responding).

Funding and Disclosures

Supported by AA024109 (FWH) from the National Institute on Alcohol and Alcoholism. The authors declare no conflicts of interest, financial or otherwise. Raw data is available from communicating author upon request.

References

1. Rudisch DM, Krasko MN, Nisbet AF et al (2022) Assays of tongue force, timing, and dynamics in rat and mouse models. Brain Res Bull 185:49–55
2. CDC (2014) Excessive drinking costs U.S. $223.5 billion. Center for Disease Control, Atlanta
3. Bouchery EE, Harwood HJ, Sacks JJ et al (2011) Economic costs of excessive alcohol consumption in the U.S., 2006. Am J Prev Med 41:516–524
4. Sacks JJ, Roeber J, Bouchery EE et al (2013) State costs of excessive alcohol consumption, 2006. Am J Prev Med 45:474–485
5. SAMHSA (2014) Risk and protective factors and initiation of substance use: results from the 2014 national survey on drug use and health. Substance Abuse and Mental Health Services Administration
6. WHO (2014) Global status report on alcohol and health-2014. World Health Organization
7. Dawson DA, Grant BF, Li TK (2005) Quantifying the risks associated with exceeding recommended drinking limits. Alcohol Clin Exp Res 29:902–908
8. Moos RH, Moos BS (2006) Rates and predictors of relapse after natural and treated remission from alcohol use disorders. Addiction 101:212–222
9. Rehm J, Mathers C, Popova S et al (2009) Global burden of disease and injury and economic cost attributable to alcohol use and alcohol-use disorders. Lancet 373:2223–2233
10. Grant BF, Chou SP, Saha TD et al (2017) Prevalence of 12-month alcohol use, high-risk drinking, and DSM-IV alcohol use disorder in the United States, 2001-2002 to 2012-2013: results from the national epidemiologic survey on alcohol and related conditions. JAMA Psychiatry 74:911–923
11. Carvalho AF, Heilig M, Perez A et al (2019) Alcohol use disorders. Lancet 394:781–792
12. White A, Castle IJ, Chen CM et al (2015) Converging patterns of alcohol use and related outcomes among females and males in the United States, 2002 to 2012. Alcohol Clin Exp Res 39:1712–1726
13. Erol A, Karpyak VM (2015) Sex and gender-related differences in alcohol use and its

consequences: contemporary knowledge and future research considerations. Drug Alcohol Depend 156:1–13

14. Brady KT, Randall CL (1999) Gender differences in substance use disorders. Psychiatr Clin North Am 22:241–252
15. Becker JB, Koob GF (2016) Sex differences in animal models: focus on addiction. Pharmacol Rev 68:242–263
16. Larimer ME, Palmer RS, Marlatt GA (1999) Relapse prevention. An overview of Marlatt's cognitive-behavioral model. Alcohol Res Health 23:151–160
17. Anton RF (2000) Obsessive-compulsive aspects of craving: development of the obsessive compulsive drinking scale. Addiction 95 (Suppl 2):S211–S217
18. Sinha R (2009) Modeling stress and drug craving in the laboratory: implications for addiction treatment development. Addict Biol 14:84–98
19. Tiffany ST, Conklin CA (2000) A cognitive processing model of alcohol craving and compulsive alcohol use. Addiction 95(Suppl 2): S145–S153
20. Epstein DH, Kowalczyk WJ (2018) Compulsive seekers: our take. Two clinicians' perspective on a new animal model of addiction. Neuropsychopharmacology 43:677–679
21. Giuliano C, Pena-Oliver Y, Goodlett CR et al (2018) Evidence for a long-lasting compulsive alcohol seeking phenotype in rats. Neuropsychopharmacology 43:728–738
22. Everitt BJ, Robbins TW (2016) Drug addiction: updating actions to habits to compulsions ten years on. Annu Rev Psychol 67:23–50
23. Hopf FW (2017) Do specific NMDA receptor subunits act as gateways for addictive behaviors? Genes Brain Behav 16:118–138
24. Hopf FW, Lesscher HM (2014) Rodent models for compulsive alcohol intake. Alcohol 48: 253–264
25. Koob GF, Volkow ND (2010) Neurocircuitry of addiction. Neuropsychopharmacology 35: 217–238
26. Kranzler HR, Soyka M (2018) Diagnosis and pharmacotherapy of alcohol use disorder: a review. JAMA 320:815–824
27. Spanagel R (2009) Alcoholism: a systems approach from molecular physiology to addictive behavior. Physiol Rev 89:649–705
28. Mann K, Roos CR, Hoffmann S et al (2018) Precision medicine in alcohol dependence: a controlled trial testing pharmacotherapy response among reward and relief drinking phenotypes. Neuropsychopharmacology 43: 891–899
29. Flores-Bonilla A, Richardson HN (2020) Sex differences in the neurobiology of alcohol use disorder. Alcohol Res 40:04
30. Guinle MIB, Sinha R (2020) The role of stress, trauma, and negative affect in alcohol misuse and alcohol use disorder in women. Alcohol Res 40:05
31. Altemus M, Sarvaiya N, Neill Epperson C (2014) Sex differences in anxiety and depression clinical perspectives. Front Neuroendocrinol 35:320–330
32. Filkowski MM, Olsen RM, Duda B et al (2017) Sex differences in emotional perception: meta analysis of divergent activation. NeuroImage 147:925–933
33. Centanni SW, Janes AC, Haggerty DL et al (2021) Better living through understanding the insula: why subregions can make all the difference. Neuropharmacology 198:108765
34. Seeley WW, Menon V, Schatzberg AF et al (2007) Dissociable intrinsic connectivity networks for salience processing and executive control. J Neurosci 27:2349–2356
35. Sridharan D, Levitin DJ, Menon V (2008) A critical role for the right fronto-insular cortex in switching between central-executive and default-mode networks. Proc Natl Acad Sci USA 105:12569–12574
36. Craig AD (2010) Once an island, now the focus of attention. Brain Struct Funct 214: 395–396
37. Menon V, Uddin LQ (2010) Saliency, switching, attention and control: a network model of insula function. Brain Struct Funct 214: 655–667
38. Kolling N, Scholl J, Chekroud A et al (2018) Prospection, perseverance, and insight in sequential behavior. Neuron 99:1069–1082 e7
39. Namkung H, Kim SH, Sawa A (2018) The insula: an underestimated brain area in clinical neuroscience, psychiatry, and neurology: (trends in neuroscience 40, 200-207, 2017). Trends Neurosci 41:551–554
40. Ochi R, Fujita N, Goto N et al (2021) Medial prefrontal area reductions, altered expressions of cholecystokinin, parvalbumin, and activating transcription factor 4 in the corticolimbic system, and altered emotional behavior in a progressive rat model of type 2 diabetes. PLoS One 16:e0256655
41. Yamashita PS, Spiacci A Jr, Hassel JE Jr et al (2017) Disinhibition of the rat prelimbic cortex promotes serotonergic activation of the dorsal

raphe nucleus and panicolytic-like behavioral effects. J Psychopharmacol 31:704–714
42. Seif T, Chang SJ, Simms JA et al (2013) Cortical activation of accumbens hyperpolarization-active NMDARs mediates aversion-resistant alcohol intake. Nat Neurosci 16:1094–1100
43. De Oliveira Sergio T, Lei K, Kwok C et al (2021) The role of anterior insula-brainstem projections and alpha-1 noradrenergic receptors for compulsion-like and alcohol-only drinking. Neuropsychopharmacology. 35: 1751–1760
44. Chen NM, Lasek AM (2020) Perineuronal nets in the insula regulate aversion-resistant alcohol drinking. Addict Biol. 25:e12821
45. Arcurio LR, Finn PR, James TW (2015) Neural mechanisms of high-risk decisions-to-drink in alcohol-dependent women. Addict Biol 20:390–406
46. Grodin EN, Sussman L, Sundby K et al (2018) Neural correlates of compulsive alcohol seeking in heavy drinkers. Biol Psych Cogn Neuro Neuroimag 2:1022–1031
47. Myrick H, Anton RF, Li X et al (2004) Differential brain activity in alcoholics and social drinkers to alcohol cues: relationship to craving. Neuropsychopharmacology 29:393–402
48. Claus ED, Ewing SW, Filbey FM et al (2011) Identifying neurobiological phenotypes associated with alcohol use disorder severity. Neuropsychopharmacology 36:2086–2096
49. Chester DS, Lynam DR, Milich R et al (2016) How do negative emotions impair self-control? A neural model of negative urgency. NeuroImage 132:43–50
50. Seo D, Jia Z, Lacadie CM et al (2011) Sex differences in neural responses to stress and alcohol context cues. Hum Brain Mapp 32: 1998–2013
51. Dager AD, Anderson BM, Rosen R et al (2014) Functional magnetic resonance imaging (fMRI) response to alcohol pictures predicts subsequent transition to heavy drinking in college students. Addiction 109:585–595
52. Lim AC, Cservenka A, Ray LA (2017) Effects of alcohol dependence severity on neural correlates of delay discounting. Alcohol Alcohol 52:506–515
53. Weafer J, Gorka SM, Dzemidzic M et al (2021) Neural correlates of inhibitory control are associated with stimulant-like effects of alcohol. Neuropsychopharmacology 46: 1442–1450
54. Srivastava AB, Sanchez-Pena J, Levin FR et al (2021) Drinking reduction during cognitive behavioral therapy for alcohol use disorder is associated with a reduction in anterior insula-bed nucleus of the stria terminalis resting-state functional connectivity. Alcohol Clin Exp Res 45:1596–1606
55. Morley KC, Logge WB, Fraser I et al (2021) High-dose baclofen attenuates insula activation during anticipatory anxiety in treatment-seeking alcohol dependant individuals: preliminary findings from a pharmaco-fMRI study. Eur Neuropsychopharmacol 46:28–36
56. Wegner SA, Hu B, De Oliveira Sergio T et al (2019) A novel NMDA receptor-based intervention to suppress compulsion-like alcohol drinking. Neuropharmacology 157:107681
57. Radke AK, Sneddon EA, Frasier RM et al (2021) Recent perspectives on sex differences in compulsion-like and binge alcohol drinking. Int J Mol Sci 22
58. McGuire JT, Botvinick MM (2010) Prefrontal cortex, cognitive control, and the registration of decision costs. Proc Natl Acad Sci USA 107:7922–7926
59. Jansma JM, Ramsey NF, de Zwart JA et al (2007) fMRI study of effort and information processing in a working memory task. Hum Brain Mapp 28:431–440
60. Hayden BY, Pearson JM, Platt ML (2011) Neuronal basis of sequential foraging decisions in a patchy environment. Nat Neurosci 14:933–939
61. Sterzer P, Kleinschmidt A (2010) Anterior insula activations in perceptual paradigms: often observed but barely understood. Brain Struct Funct 214:611–622
62. Hillman KL, Bilkey DK (2013) Persisting through subjective effort: a key role for the anterior cingulate cortex? Behav Brain Sci 36(691–2):707–726
63. Kolling N, Wittmann M, Rushworth MFS (2014) Multiple neural mechanisms of decision making and their competition under changing risk pressure. Neuron 81:1190–1202
64. Daniel ML, Cocker PJ, Lacoste J et al (2017) The anterior insula bidirectionally modulates cost-benefit decision-making on a rodent gambling task. Eur J Neurosci 46:2620–2628
65. Wang Y, Liu Y, Yang L et al (2015) Novelty seeking is related to individual risk preference and brain activation associated with risk prediction during decision making. Sci Rep 5: 10534
66. Jenni NL, Larkin JD, Floresco SB (2017) Prefrontal dopamine D1 and D2 receptors regulate dissociable aspects of decision making via distinct ventral striatal and Amygdalar circuits. J Neurosci 37:6200–6213
67. Meder D, Haagensen BN, Hulme O et al (2016) Tuning the brake while raising the

stake: network dynamics during sequential decision-making. J Neurosci 36:5417–5426

68. van der Laan LN, de Ridder DT, Charbonnier L et al (2014) Sweet lies: neural, visual, and behavioral measures reveal a lack of self-control conflict during food choice in weight-concerned women. Front Behav Neurosci 8:184
69. Frankenstein UN, Richter W, McIntyre MC et al (2001) Distraction modulates anterior cingulate gyrus activations during the cold pressor test. NeuroImage 14:827–836
70. Zeng Y, Hu D, Yang W et al (2018) A voxel-based analysis of neurobiological mechanisms in placebo analgesia in rats. NeuroImage 178: 602–612
71. Kaplan JT, Gimbel SI, Harris S (2016) Neural correlates of maintaining one's political beliefs in the face of counterevidence. Sci Rep 6: 39589
72. Aimone JA, Houser D, Weber B (2014) Neural signatures of betrayal aversion: an fMRI study of trust. Proc Biol Sci 281:20132127
73. Benoit RG, Gilbert SJ, Burgess PW (2011) A neural mechanism mediating the impact of episodic prospection on farsighted decisions. J Neurosci 31:6771–6779
74. Nelson SM, Dosenbach NU, Cohen AL et al (2010) Role of the anterior insula in task-level control and focal attention. Brain Struct Funct 214:669–680
75. Stoppel CM, Boehler CN, Strumpf H et al (2011) Neural processing of reward magnitude under varying attentional demands. Brain Res 1383:218–229
76. Grupe DW, Oathes DJ, Nitschke JB (2013) Dissecting the anticipation of aversion reveals dissociable neural networks. Cereb Cortex 23: 1874–1883
77. Sadaghiani S, D'Esposito M (2015) Functional characterization of the Cingulo-Opercular network in the maintenance of tonic alertness. Cereb Cortex 25:2763–2773
78. Darevsky D, Gill TM, Vitale KR et al (2019) Drinking despite adversity: behavioral evidence for a head down and push strategy of conflict-resistant alcohol drinking in rats. Addict Biol 24:426–437
79. Darevsky D, Hopf FW (2020) Behavioral indicators of succeeding and failing under higher-challenge compulsion-like alcohol drinking in rat. Behav Brain Res 393:112768
80. De Oliveira Sergio T, Darevsky D, de Paula Soares V, de Cassia Albino M, Maulucci D, Wean S, Hopf FW Evidence for different greater-persistence strategies under lower and higher challenge for alcohol in female rats. Biorvix:www.biorxiv.org/content/10.1101/2022.05.18.492488v1
81. Starski P, Morningstar M, Katner S, Frasier R, De Oliveira Sergio T, Wean S, Lapish C, Hopf FW Anterior insular cortex firing links initial and sustained encoding during aversion-resistant alcohol consumption. Biorvix. https://doi.org/10.1101/2022.05.24.493243
82. Ottenheimer D, Richard JM, Janak PH (2018) Ventral pallidum encodes relative reward value earlier and more robustly than nucleus accumbens. Nat Commun 9:4350
83. Bari BA, Grossman CD, Lubin EE et al (2019) Stable representations of decision variables for flexible behavior. Neuron 103:922–933 e7
84. Mukherjee N, Wachutka J, Katz DB (2019) Impact of precisely-timed inhibition of gustatory cortex on taste behavior depends on single-trial ensemble dynamics. elife 8:e45968
85. Sadacca BF, Mukherjee N, Vladusich T et al (2016) The behavioral relevance of cortical neural ensemble responses emerges suddenly. J Neurosci 36:655–669
86. Roebber JK, Izenwasser S, Chaudhari N (2015) Cocaine decreases saccharin preference without altering sweet taste sensitivity. Pharmacol Biochem Behav 133:18–24
87. Moore JD, Kleinfeld D, Wang F (2014) How the brainstem controls orofacial behaviors comprised of rhythmic actions. Trends Neurosci 37:370–380
88. Seif T, Simms JA, Lei K et al (2015) D-Serine and D-Cycloserine reduce compulsive alcohol intake in rats. Neuropsychopharmacology 40: 2357–2367
89. Davis JD, Smith GP (1992) Analysis of the microstructure of the rhythmic tongue movements of rats ingesting maltose and sucrose solutions. Behav Neurosci 106:217–228
90. Patwell R, Yang H, Pandey SC et al (2021) An operant ethanol self-administration paradigm that discriminates between appetitive and consummatory behaviors reveals distinct behavioral phenotypes in commonly used rat strains. Neuropharmacology 201:108836
91. Lardeux S, Kim JJ, Nicola SM (2013) Intermittent access to sweet high-fat liquid induces increased palatability and motivation to consume in a rat model of binge consumption. Physiol Behav 114–115:21–31
92. Spector AC, Klumpp PA, Kaplan JM (1998) Analytical issues in the evaluation of food deprivation and sucrose concentration effects

on the microstructure of licking behavior in the rat. Behav Neurosci 112:678–694

93. Naneix F, Peters KZ, McCutcheon JE (2020) Investigating the effect of physiological need states on palatability and motivation using microstructural analysis of licking. Neuroscience 447:155–166
94. Samson HH (2000) The microstructure of ethanol drinking: genetic and behavioral factors in the control of drinking patterns. Addiction 95(Suppl 2):S61–S72
95. Barkley-Levenson AM, Crabbe JC (2015) Distinct ethanol drinking microstructures in two replicate lines of mice selected for drinking to intoxication. Genes Brain Behav 14: 398–410
96. Haggerty DL, Munoz B, Pennington T et al (2022) The role of anterior insular cortex inputs to dorsolateral striatum in binge alcohol drinking. elife 11:e77411
97. Davis JD, Perez MC (1993) Food deprivation- and palatability-induced microstructural changes in ingestive behavior. Am J Phys 264: R97–R103
98. Hopf FW, Chang SJ, Sparta DR et al (2010) Motivation for alcohol becomes resistant to quinine adulteration after 3 to 4 months of intermittent alcohol self-administration. Alcohol Clin Exp Res 34:1565–1573
99. Hsiao S, Fan RJ (1993) Additivity of taste-specific effects of sucrose and quinine: microstructural analysis of ingestive behavior in rats. Behav Neurosci 107:317–326
100. Baird JP, St John SJ, Nguyen EA (2005) Temporal and qualitative dynamics of conditioned taste aversion processing: combined generalization testing and licking microstructure analysis. Behav Neurosci 119: 983–1003
101. Renteria R, Cazares C, Gremel CM (2020) Habitual ethanol seeking and licking microstructure of enhanced ethanol self-administration in ethanol-dependent mice. Alcohol Clin Exp Res 44:880–891
102. Flores-Bonilla A, De Oliveira B, Silva-Gotay A et al (2021) Shortening time for access to alcohol drives up front-loading behavior, bringing consumption in male rats to the level of females. Biol Sex Differ 12:51
103. Baker EJ, Farro J, Gonzales S et al (2014) Chronic alcohol self-administration in monkeys shows long-term quantity/frequency categorical stability. Alcohol Clin Exp Res 38:2835–2843
104. Gowin JL, Sloan ME, Stangl BL et al (2017) Vulnerability for alcohol use disorder and rate of alcohol consumption. Am J Psychiatry 174: 1094–1101
105. Robinson SL, McCool BA (2015) Microstructural analysis of rat ethanol and water drinking patterns using a modified operant self-administration model. Physiol Behav 149:119–130
106. Parent MA, Amarante LM, Swanson K et al (2015b) Cholinergic and ghrelinergic receptors and KCNQ channels in the medial PFC regulate the expression of palatability. Front Behav Neurosci 9:284
107. Patton MS, Heckman M, Kim C et al (2021) Compulsive alcohol consumption is regulated by dorsal striatum fast-spiking interneurons. Neuropsychopharmacology 46:351–359
108. Gutierrez R, Carmena JM, Nicolelis MA et al (2006) Orbitofrontal ensemble activity monitors licking and distinguishes among natural rewards. J Neurophysiol 95:119–133
109. Baird JP, Choe A, Loveland JL et al (2009) Orexin-a hyperphagia: hindbrain participation in consummatory feeding responses. Endocrinology 150:1202–1216
110. Voon V, Derbyshire K, Ruck C et al (2015) Disorders of compulsivity: a common bias towards learning habits. Mol Psychiatry 20: 345–352
111. Ersche KD, Lim TV, Ward LHE et al (2017) Creature of habit: a self-report measure of habitual routines and automatic tendencies in everyday life. Pers Individ Dif 116:73–85
112. Han SW, Eaton HP, Marois R (2019) Functional fractionation of the Cingulo-Opercular network: alerting insula and updating cingulate. Cereb Cortex 29:2624–2638
113. Craig AD (2009) How do you feel – now? The anterior insula and human awareness. Nat Rev Neurosci 10:59–70
114. Jaramillo AA, Randall PA, Stewart S et al (2018b) Functional role for cortical-striatal circuitry in modulating alcohol self-administration. Neuropharmacology 130: 42–53
115. Dao NC, Brockway DF, Suresh Nair M et al (2021) Somatostatin neurons control an alcohol binge drinking prelimbic microcircuit in mice. Neuropsychopharmacology 46:1906–1917
116. Timme NM, Ma B, Linsenbardt D et al (2022) Compulsive alcohol drinking in rodents is associated with altered representations of behavioral control and seeking in dorsal medial prefrontal cortex. Nat Commun 13:3990
117. Hamilton DA, Barto D, Rodriguez CI et al (2014) Effects of moderate prenatal ethanol

exposure and age on social behavior, spatial response perseveration errors and motor behavior. Behav Brain Res 269:44–54
118. Whishaw IQ, Tompkins GJ (1988) An optic-fiber photocell detector for measuring tongue protrusion in the rat: evaluation of recovery from localized cortical lesions. Physiol Behav 43:397–401
119. Whishaw IQ, Kolb B (1983) "Stick out your tongue": tongue protrusion in neocortex and hypothalamic damaged rats. Physiol Behav 30:471–480
120. Castro AJ (1972) The effects of cortical ablations on tongue usage in the rat. Brain Res 45: 251–253
121. Brimley CC, Mogenson GJ (1979) Oral motor deficits following lesions of the central nervous system in the rat. Am J Phys 237: R126–R131
122. Shipley JE, Rowland N, Antelman SM (1980) Orbital or medial frontal cortical lesions have different effects on tail pressure-elicited oral behaviors in rats. Physiol Behav 24:1091–1094
123. Parent MA, Amarante LM, Liu B et al (2015a) The medial prefrontal cortex is crucial for the maintenance of persistent licking and the expression of incentive contrast. Front Integr Neurosci 9:23
124. Horst NK, Laubach M (2009) The role of rat dorsomedial prefrontal cortex in spatial working memory. Neuroscience 164:444–456
125. Narayanan NS, Horst NK, Laubach M (2006) Reversible inactivations of rat medial prefrontal cortex impair the ability to wait for a stimulus. Neuroscience 139:865–876
126. Krigolson O, Bell J, Kent CM et al (2012) Reduced cortical motor potentials underlie reductions in memory-guided reaching performance. Mot Control 16:353–370
127. Smith NJ, Horst NK, Liu B et al (2010) Reversible inactivation of rat premotor cortex impairs temporal preparation, but not inhibitory control, during simple reaction-time performance. Front Integr Neurosci 4:124
128. Afonso VM, Sison M, Lovic V et al (2007) Medial prefrontal cortex lesions in the female rat affect sexual and maternal behavior and their sequential organization. Behav Neurosci 121:515–526
129. Horst NK, Laubach M (2013) Reward-related activity in the medial prefrontal cortex is driven by consumption. Front Neurosci 7:56
130. Stuss DT, Murphy KJ, Binns MA et al (2003) Staying on the job: the frontal lobes control individual performance variability. Brain 126: 2363–2380
131. Stuss DT, Alexander MP, Shallice T et al (2005) Multiple frontal systems controlling response speed. Neuropsychologia 43:396–417
132. Croxson PL, Walton ME, O'Reilly JX et al (2009) Effort-based cost-benefit valuation and the human brain. J Neurosci 29:4531–4541
133. Simon NW, Wood J, Moghaddam B (2015) Action-outcome relationships are represented differently by medial prefrontal and orbitofrontal cortex neurons during action execution. J Neurophysiol 114:3374–3385
134. Rudebeck PH, Walton ME, Smyth AN et al (2006) Separate neural pathways process different decision costs. Nat Neurosci 9:1161–1168
135. Rich EL, Wallis JD (2016) Decoding subjective decisions from orbitofrontal cortex. Nat Neurosci 19:973–980
136. Sharpe MJ, Killcross S (2015) The prelimbic cortex directs attention toward predictive cues during fear learning. Learn Mem 22: 289–293
137. Park J, Moghaddam B (2017) Impact of anxiety on prefrontal cortex encoding of cognitive flexibility. Neuroscience 345:193–202
138. Park J, Wood J, Bondi C et al (2016) Anxiety evokes hypofrontality and disrupts rule-relevant encoding by dorsomedial prefrontal cortex neurons. J Neurosci 36:3322–3335
139. Uddin LQ, Nomi JS, Hebert-Seropian B et al (2017) Structure and function of the human insula. J Clin Neurophysiol 34:300–306
140. Hayes DJ, Duncan NW, Xu J et al (2014) A comparison of neural responses to appetitive and aversive stimuli in humans and other mammals. Neurosci Biobehav Rev 45:350–368
141. Hauser TU, Eldar E, Dolan RJ (2017) Separate mesocortical and mesolimbic pathways encode effort and reward learning signals. Proc Natl Acad Sci USA 114:E7395–E7404
142. Ryman SG, El Shaikh AA, Shaff NA et al (2018) Proactive and reactive cognitive control rely on flexible use of the ventrolateral prefrontal cortex. Hum Brain Mapp 40:955–966
143. Hardung S, Epple R, Jackel Z et al (2017) A functional gradient in the rodent prefrontal cortex supports behavioral inhibition. Curr Biol 27:549–555

144. Chua HF, Gonzalez R, Taylor SF et al (2009) Decision-related loss: regret and disappointment. NeuroImage 47:2031–2040
145. Burke CJ, Tobler PN (2011) Reward skewness coding in the insula independent of probability and loss. J Neurophysiol 106: 2415–2422
146. Kolling N, Behrens TE, Mars RB et al (2012) Neural mechanisms of foraging. Science 336: 95–98
147. Dosenbach NU, Visscher KM, Palmer ED et al (2006) A core system for the implementation of task sets. Neuron 50:799–812
148. Wittmann MK, Fouragnan E, Folloni D et al (2020) Global reward state affects learning and activity in raphe nucleus and anterior insula in monkeys. Nat Commun 11:3771
149. Guillem K, Kravitz AV, Moorman DE et al (2010) Orbitofrontal and insular cortex: neural responses to cocaine-associated cues and cocaine self-administration. Synapse 64:1–13
150. DeCoteau WE, Kesner RP, Williams JM (1997) Short-term memory for food reward magnitude: the role of the prefrontal cortex. Behav Brain Res 88:239–249
151. Pribut HJ, Vazquez D, Brockett AT et al (2021) Prior cocaine exposure increases firing to immediate reward while attenuating cue and context signals related to reward value in the insula. J Neurosci 41:4667–4677
152. Laubach M, Caetano MS, Narayanan NS (2015) Mistakes were made: neural mechanisms for the adaptive control of action initiation by the medial prefrontal cortex. J Physiol 109:104–117
153. Ito S, Ogawa H (1994) Neural activities in the fronto-opercular cortex of macaque monkeys during tasting and mastication. Jpn J Physiol 44:141–156
154. Petyko Z, Galosi R, Toth A et al (2015) Responses of rat medial prefrontal cortical neurons to Pavlovian conditioned stimuli and to delivery of appetitive reward. Behav Brain Res 287:109–119
155. Amarante LM, Caetano MS, Laubach M (2017) Medial frontal theta is entrained to rewarded actions. J Neurosci 37:10757–10769
156. Yaxley S, Rolls ET, Sienkiewicz ZJ (1988) The responsiveness of neurons in the insular gustatory cortex of the macaque monkey is independent of hunger. Physiol Behav 42: 223–229
157. Hollander JA, Lu Q, Cameron MD et al (2008) Insular hypocretin transmission regulates nicotine reward. Proc Natl Acad Sci USA 105:19480–19485
158. Pelloux Y, Murray JE, Everitt BJ (2013) Differential roles of the prefrontal cortical subregions and basolateral amygdala in compulsive cocaine seeking and relapse after voluntary abstinence in rats. Eur J Neurosci 38:3018–3026
159. Rotge JY, Cocker PJ, Daniel ML et al (2017) Bidirectional regulation over the development and expression of loss of control over cocaine intake by the anterior insula. Psychopharmacology 234:1623–1631
160. Baldo BA, Spencer RC, Sadeghian K et al (2016) GABA-mediated inactivation of medial prefrontal and agranular insular cortex in the rat: contrasting effects on hunger- and palatability-driven feeding. Neuropsychopharmacology 41:960–970
161. Pushparaj A, Le Foll B (2015) Involvement of the caudal granular insular cortex in alcohol self-administration in rats. Behav Brain Res 293:203–207
162. Jasmin L, Burkey AR, Granato A et al (2004) Rostral agranular insular cortex and pain areas of the central nervous system: a tract-tracing study in the rat. J Comp Neurol 468:425–440
163. Halcomb ME, Chumin EJ, Goni J et al (2019) Aberrations of anterior insular cortex functional connectivity in nontreatment-seeking alcoholics. Psychiatry Res Neuroimaging 284:21–28
164. Tan Y, Yan R, Gao Y et al (2022) Spatial-topographic nestedness of interoceptive regions within the networks of decision making and emotion regulation: combining ALE meta-analysis and MACM analysis. NeuroImage 260:119500
165. Porges EC, Decety J (2013) Violence as a source of pleasure or displeasure is associated with specific functional connectivity with the nucleus accumbens. Front Hum Neurosci 7: 447
166. Poeppl TB, Nitschke J, Dombert B et al (2011) Functional cortical and subcortical abnormalities in pedophilia: a combined study using a choice reaction time task and fMRI. J Sex Med 8:1660–1674
167. Kesner RP, Gilbert PE (2007) The role of the agranular insular cortex in anticipation of reward contrast. Neurobiol Learn Mem 88: 82–86
168. Smith KS, Graybiel AM (2013) A dual operator view of habitual behavior reflecting cortical and striatal dynamics. Neuron 79:361–374

169. Jezzini A, Caruana F, Stoianov I et al (2012) Functional organization of the insula and inner perisylvian regions. Proc Natl Acad Sci USA 109:10077–10082
170. Stapleton JR, Lavine ML, Wolpert RL et al (2006) Rapid taste responses in the gustatory cortex during licking. J Neurosci 26:4126–4138
171. Haaranen M, Scuppa G, Tambalo S et al (2020b) Anterior insula stimulation suppresses appetitive behavior while inducing forebrain activation in alcohol-preferring rats. Transl Psychiatry 10:150
172. Pushparaj A, Kim AS, Musiol M et al (2015) Differential involvement of the agranular vs granular insular cortex in the acquisition and performance of choice behavior in a rodent gambling task. Neuropsychopharmacology 40:2832–2842
173. Arguello AA, Wang R, Lyons CM et al (2017) Role of the agranular insular cortex in contextual control over cocaine-seeking behavior in rats. Psychopharmacology 234:2431–2441
174. Mendez-Ruette M, Linsambarth S, Moraga-Amaro R et al (2019) The role of the rodent insula in anxiety. Front Physiol 10:330

Chapter 5

Applying Functional MRI to Study Adolescent Substance Use

Kathleen I. Crum, Kristy Bixler, and Leslie A. Hulvershorn

Abstract

Functional magnetic resonance imaging is a tool used to examine brain activity both at rest and in response to a variety of stimuli or tasks. Functional MRI may be applied to study correlates of substance use and risk for substance use in adolescence. In this chapter, we describe design, logistical, and analytic considerations for conducting a functional MRI research study among adolescents engaged in or at risk for substance use. Our objective is to provide readers with an introductory guide to applying functional MRI in this specialized population, as well as highlight potential resources to obtain further information.

Key words Functional MRI, Adolescents, Substance use disorder, Substance use risk behaviors, Neuroimaging experimental design

1 Introduction: Adolescent Substance Use

Initiating substance use in adolescence is associated with a long-term trajectory of impairment, including higher rates of polysubstance use, substance use disorder (SUD), and other psychiatric disorders [1]. Initiation of substance use during adolescence is associated with lower treatment recovery rates and lower academic and occupational achievement [1]. Further, substance use in adolescence is associated with long-term health risks, both directly related to effects of substances on the brain and body and indirectly through effects on other risk behaviors [2]. Early- to late-adolescence (ages 11–17) represents a sensitive period for initiation and escalation of substance use [1]. Research on preventing adolescent substance use has identified multiple risk and protective factors for initiation and escalation of substance use, as well as development of SUD. A major class of risk factors involves brain response to rewards, emotions, and decision-making.

Melissa A. Cyders (ed.), *Translational Research Methods for Alcohol Use Disorders*, Neuromethods, vol. 201,
https://doi.org/10.1007/978-1-0716-3267-3_5,

2 Functional Magnetic Resonance Imaging (fMRI) as a Measurement Tool in Risk for, and Correlates of, Adolescent Substance Use

Brain response to a variety of stimuli can be measured using functional magnetic resonance imaging (fMRI) [3, 4]. fMRI is a non-invasive imaging technique that indexes brain activity by measuring changes associated with blood flow [5] during a session in an MRI scanner. This technique involves manipulating hydrogen nuclei by applying a magnetic pulse at a specific radio frequency, which generates a magnetic resonance signal that is deleted by specialized coils in the scanner. The signal can be used to generate an image [5] mapping the spatial distribution of the signal. MRI can be applied to mapping both brain anatomy or structure, and brain function. The focus of this chapter is on the use of MRI to approximate brain *function* in response to stimuli and situations that are associated with adolescent substance use and related sequelae.

Neuronal activity corresponds to changes in the magnetic resonance signal, which is sensitive to changes in blood flow that accompany changes in neuronal activity. The resulting fMRI data can be analyzed to identify brain regions in which changes in the signal correspond with changes in a given stimulus (e.g., viewing a fearful face, winning a monetary reward, choosing between two behavioral options), or changes in spontaneous brain activity at rest. The level of oxygen in blood is related to its magnetic properties because of oxygen–hemoglobin binding in the blood. The magnetic resonance signal is stronger in the presence of more highly oxygenated blood. Neuronal activity is associated with a significant increase in oxygenated blood flow. This phenomenon gives rise to the blood oxygenation level-dependent (BOLD) data used in fMRI analyses [5]. fMRI allows localization of brain activity so that responses to a given stimulus can be mapped to brain regions and large-scale brain networks [6]. Metrics that can be derived include not only BOLD data but also psychophysiological interaction (PPI) [7] data. PPI data represent functional connectivity, or correlations between activation in one brain region and another.

One type of fMRI measurement—task-based fMRI—involves collecting data while individuals are performing tasks in the MRI scanner. During task-based fMRI data analysis, brain activation data are paired with stimulus timings and behavioral responses. This pairing allows researchers to associate brain activation with a particular behavioral task. Both BOLD data and functional connectivity data may be used to estimate these associations.

Another type of fMRI measurement—resting state fMRI—involves collecting data while research participants are awake but not engaged in a task in the MRI scanner. Resting state fMRI typically involves focusing on a crosshair stimulus on an otherwise-blank screen, without additional instructions or task demands. Inferences drawn from resting state fMRI are based on spontaneous BOLD signal fluctuations—rather than fluctuations in relation to stimulus presentations and task demands—which are thought to represent spontaneous brain activation. During resting state fMRI data analysis, brain activation data may be analyzed as functional connectivity *between* one brain region or network and another, or *within* a single brain region or network [8]. Multiple resting state fMRI metrics [9, 10] exist. Graph theory may be used to assess global and local connectivity. Seed-based connectivity may be used to assess connectivity between a specific brain area—the "seed" region—and other brain areas. Regional homogeneity and amplitude of low frequency fluctuations may be used to assess local connectivity.

3 Design Considerations: Choosing Tasks

Prudent experimental design is critical to conducting an fMRI study that will yield interpretable, informative findings. One aspect of experimental design concerns choosing appropriate tasks for task-based fMRI. Broadly, tasks should be reliable and valid among youth in the age range of interest, and for the topic of interest. We will briefly discuss representative tasks that have demonstrated sound psychometric properties for problematic substance use and risk for substance use in adolescents ages 11–17. Tasks fall under the following broad domains: reward response and reinforcement learning, threat response and emotion regulation, cue reactivity, and response inhibition. Table 1 provides popular task examples within each domain (by no means an exhaustive list).

Several tasks of interest are included in the ongoing Adolescent Behavioral and Cognitive Development project [11] ($N = 11{,}875$) [12], a surveillance study following youth across development. One of the goals of the ABCD study is to identify risk and protective factors for early substance use and SUD trajectories. The Emotional N-back task is designed to activate brain regions involved in emotional responding—including threat response—as well as brain regions involved in implicit emotion regulation [13, 14]. The implicit emotion regulation aspect of the task involves completing a working memory task after exposure to cues of varying emotional valences. The Monetary Incentive Delay (MID) task is designed to activate brain regions involved in reward responding and motivation; in this task, youth must respond quickly enough to a probe stimulus to avoid win money or losing money [15]. The Stop Signal task is designed to activate brain regions involved in response inhibition; in this task, youth are presented with a cue to proceed with a motor response, but must interrupt that motor response

Table 1
Commonly used tasks in functional MRI studies examining adolescents with, or at risk for, substance use disorders

Task	Domain	References
Doors	Cue reactivity	Crane et al. [21] and Carlson et al. [22]
Emotional N-back	Threat response and emotion regulation	Cohen et al. [23] and Barch et al. [13]
Affective Stroop	Threat response and emotion regulation	Blair et al. [24] and Aloi et al. [25]
Facial Morphs	Threat response and emotion regulation	Makrushin et al. [26] and Windhager et al. [27]
Go-No Go	Response inhibition	Roberts et al. [28] and Hare et al. [29]
Stop Signal	Response inhibition	Logan [16] and Verbruggen and Logan [30]
Balloon Analog Risk Task (BART)	Reward and reinforcement learning	Lejuez et al. [18] and Hulvershorn et al. [19]
Delay Discounting	Reward and reinforcement learning	Butcher et al. [20] and Lempert et al. [31]
Monetary Incentive Delay (MID)	Reward and reinforcement learning	Knutson et al. [15] and Balodis and Potenza [32]
Passive Avoidance	Reward and reinforcement learning	Aloi et al. [33] and Schlund et al. [34]
Wheel of Fortune	Reward and reinforcement learning	Ernst et al. [35], Rao et al. [36], and Shad et al. [37]

when it is followed by a cue to stop [16]. Casey and colleagues (2018) [17] provide a full breakdown of neural correlates associated with each task and their National Institutes of Health Collaborative Research on Addiction at NIH (CRAN) [11] domains of function.

Finally, we will discuss two tasks used by our research group to investigate brain function among young adolescents at risk for substance use, by virtue of externalizing disorders and family history of SUD. The Balloon Analog Risk Task (BART) [18, 19] and the Delay Discounting task [20] are designed to activate brain regions involved in reward and reinforcement learning, and the behavioral domain of decision-making. In the BART, youth are presented with a computerized balloon; the more the balloon is inflated, the more money participants earn—but if the balloon becomes too inflated and pops, participants lose all their money. On each trial, participants have a choice whether to inflate the balloon with an additional air "pump," or whether to collect their earnings so far. Size at which the balloon pops, and earnings are lost, varies throughout the task. Youth must decide from varied options in the context of increasing risk, that is, the amount they risk losing, and the probability that they will lose, increases with each balloon pump. In the Delay Discounting task, youth are asked to choose between an offer of immediate versus delayed monetary rewards. The size of the reward and length of the delay are parametrically modulated across trials. Youth must choose the optimal balance between these two factors on each trial.

Another aspect of experimental design concerns total scan length. Youth may have difficulty remaining still in the MRI scanner for an extended period, particularly youth with psychiatric difficulties commonly comorbid with substance use or risk for substance use—such as ADHD. Many fMRI tasks require multiple sessions ("runs") to obtain enough data points for analysis. Resting state fMRI scans must also reach a certain length (approximately 7 min) [38, 39] to acquire enough data for analysis following motion outlier removal. Task length, number of runs, and resting state scan duration should be considered and balanced with reliability and signal-to-noise ratio. In our experience, total scan length of 1 h or less is feasible for youth in middle childhood and up. Total scan length calculations should include acquisition of anatomical and functional scans, as well as field mapping images.

4 Preparing Youth Who Use Substances, and Who Are at Risk for Substance Use, for the Scanner

Precautions should be taken prior to scanning youth. Youth—especially older adolescents—are often reliable reporters of their scanning eligibility. However, collateral sources confirming eligibility should be obtained for safety. Youth report alone may be limited by lack of awareness of medical conditions or medical history, impaired judgment associated with substance use, and apprehension about potential consequences. Both youth and caregiver report should be obtained during medical screening, including presence of metal in or on the body, which is particularly important as the MRI scanner generates strong magnetic forces. Pregnancy testing is often conducted on all youth assigned a female sex at birth, although fetal MRI is becoming more widely accepted. Pregnancy testing is typically accomplished by using urine samples. State and federal laws should be considered during planning for positive pregnancy test follow-up procedures, including potential mandated safety/child abuse reporting.

Of particular relevance to youth at risk for or engaged in substance use, alcohol, and drug screening are often conducted prior to entering the scanner. Acute intoxication may affect task performance and brain activation during tasks or at rest

[40]. Therefore, drug and alcohol screening should be conducted on youth prior to scanning. Drug and alcohol screening is typically conducted using urine and breath samples, though blood tests may also be used. Individuals who are intoxicated should be rescheduled for assessments and scans and the focus of the visit should transition to ensuring that the participant can be safely transferred home or to appropriate treatment services (e.g., emergency department for severe intoxication).

Mock scanning may be helpful when scanning youth, especially youth with psychiatric conditions—as is often the case for youth at risk for or engaged in substance use [41, 42]. A mock scanner model is used to prepare youth for the space constraints and the need for being still during the actual MRI scan. The mock scanner is a scaled-down replica of an MRI scanner constructed of plastic or wood, with a sliding table, speaker system, and faux head coil similar to an actual MRI scanner. Software and remote sensors are available to engage youth in a stillness-training task; sensors can be attached to a headband of other device worn on the head, allowing youth to practice maintaining a specified head position and to receive real-time feedback. Mock scan procedures allow youth to acclimate to the close surroundings, head and body position, and stillness required in the MRI scanner, as well as the noises they will hear in the scanner.

Introducing, demonstrating, and practicing fMRI tasks in advance is important, especially when scanning youth. This is particularly relevant for youth engaged in substance use, who may be susceptible to the neurocognitive effects of substances [43]. Practice sessions are typically conducted outside the scanner, mimicking conditions as closely as possible, e.g., using keys on a keyboard that will closely match the hand motions and buttons youth will use in the MRI scanner. Youth should be provided with simple written and verbal instructions, along with corresponding illustrations to accompany varied reading proficiency and developmental levels. One to two abbreviated practice runs should be conducted per fMRI task. Hand placement and button presses should also be demonstrated inside the scanner prior to beginning acquisition, allowing youth to practice using the response devices provided and receive real-time feedback.

5 Special Considerations During Scanning

We recommend frequent communication between research staff and youth during scans. Brief, simplified instructions should be provided to youth frequently, including before acquisition begins, between each set of scans (e.g., between the anatomical and task or resting state scans), and after acquisition is complete. Researchers should consider eliciting button press responses indicating understanding of questions and instructions, and readiness to begin and transition to the next set of scans, as part of their scanning protocol. Staff running scans should monitor response patterns from the scan control room to detect erratic patterns, lengthy delays, or no response—which may indicate lack of understanding of the task, wandering attention, anxiety, or fatigue.

Feedback is also helpful in preventing and addressing motion. Managing the effects of motion is more effective at the level of data acquisition, rather than data analysis. Following the above recommendations regarding mock scanning may be helpful in reducing motion. We also recommend closely monitoring motion using in-line images during scans. Feedback should be provided to youth between sets of scans (e.g., between task runs). Extra time should be budgeted at the MRI scanner to allow repeat scans, if needed. Researchers may also consider planning for higher potential dropout rates based on motion-related issues for youth studies than they would for adult studies.

Researchers should also consider designing scan protocol such that anatomical scans are collected first. The benefits of this practice are two-fold. Most analyses *require* a corresponding anatomical scan with which to pair functional images. Further, detecting significant motion during the anatomical scan allows research staff to provide feedback on motion to youth early in the scan session.

Finally, MRI-compatible eye-tracking devices and accompanying software may be helpful in scanning youth. Eye tracking may be used to measure constructs for which eye gaze serves as a proxy (e.g., attention). However, the small faces of younger children and adolescents may present a challenge to calibrating eye-tracking devices meant to detect pupil activity. Less of the eye may be visible through certain MRI head coils; coils with a design that fits over the face may cast a shadow. Thorough trialing of eye-tracking devices and their calibration is recommended.

6 Analytic Considerations

During the study setup process and early data collection, we recommend frequent testing of the data processing and analysis "pipelines." The "pipeline" is the set of processes by which researchers convert their data from the format generated by the MRI scanner to a format that can be used to answer research questions. The pipeline broadly involves preprocessing, individual participant-level analysis, and group-level analysis. Interim preprocessing and analyses should also be conducted on a regular, frequent basis to identify and takes steps to resolve any issues with data collection. For details on these processes and setting up a pipeline, we recommend consulting Huettel, Song, and McCarthy's "Functional Magnetic Resonance Imaging" textbook [44].

Here we will discuss several issues relevant to functional MRI scanning with youth engaging in, or at risk for, substance use. First, motion may present a particular analytical challenge for youth in this special population. Substance use is often preceded and accompanied by other psychiatric issues, including conduct problems, attention problems, and anxiety. In turn, youth with these psychiatric issues may have difficulty remaining still in the scanner. Strategies to manage motion are most effective in the planning stages of a project; however, motion can be addressed during preprocessing and analyses to some extent. Common methods of addressing motion include, but are not limited to, the following: (1) motion correction algorithms employed during preprocessing, which involves aligning brain volumes to a reference scan; (2) censoring brain volumes over a specified motion threshold; (3) including motion parameters as covariates in individual-level analyses (with or without the derivatives and/or squares of these covariates); and (4) independent components analysis (ICA) [45] to identify and remove signal that likely arises from motion or other "noise." If motion is a serious issue for a particular scan session or participant, and motion significantly affects data quality, the particular scan session in question may be discarded on an individual basis. Examining data quality for each participant and scan session is highly recommended.

Second, test-retest reliability is an ongoing issue in neuroscience (as is the case for multiple constructs within psychology and psychiatry) [46]. For some constructs, we may expect change across time, development, and as a function of other factors such as pharmacologic or psychosocial treatment. However, there may be problems replicating results within the same group of participants across time—and across development. Several recommendations have been made to increase reliability in fMRI data [47, 48]. First, researchers should consider conducting repeat scans at varying intervals for a subset of participants, if possible, to evaluate reliability across scans. Second, researchers should avoid contrasts during group analysis, that is, generating contrasts by subtracting signal specific to one condition from another condition (e.g., Task 1—Task 2). Generating contrasts using this method may "average out" or mathematically remove variability and therefore reduce power to detect differences between conditions.

Third, there are two analytic standards that may be sensitive to developmental changes. One is the hemodynamic response function. The hemodynamic response function may differ across development. Specifically, studies have reported increased time-to-peak and decreased peak amplitude with increasing age [49, 50]. We recommend reviewing the literature available on specific age groups of interest during study design stages, and examining data at the preprocessing and analysis stages to determine which hemodynamic response function works best on a study-by-study basis. A second standard is the anatomical atlas or template used to identify brain regions. Brain development continues to occur across childhood and adolescence, both on a structural and functional level. Therefore, youth and adult neuroanatomies differ [51]. Different age groups may show differences in development and function of the brain over time. The rate and timing of changes depends on the brain structures and specific process involved (e.g., decision-making, stimulus value representation, reward [52–55] response). For example, hippocampal volume may increase with age, and striatal volume may decrease with age, while some subcortical structure volumes follow U-shaped curves and peak at different points in childhood and adolescence [56]. We recommend weighing the benefits of using a pediatric template versus an adult template during data analysis and interpretation (e.g., MNI vs. MNI Pediatric Templates) [57–59].

References

1. Substance Abuse and Mental Health Services Administration Center for Behavioral Health Statistics and Quality (2014) The TEDS report: age of substance use initiation among treatment admissions aged 18 to 30. Substance Abuse and Mental Health Services Administration Center, Rockville
2. Centers for Disease Control and Prevention. Fetal alcohol spectrum disorders (FASDs). https://www.cdc.gov/ncbddd/fasd/features/teen-substance-use.html. Published 2020. Accessed 4 Aug 2022
3. Gore JC (2003) Principles and practice of functional MRI of the human brain. J Clin Invest 112(1):4–9. PMC162295
4. Glover GH (2011) Overview of functional magnetic resonance imaging. Neurosurg Clin N Am 22(2):133–139, vii. PMC3073717
5. University of California San Diego School of Medicine, Center for Functional MRI. What is fMRI? http://fmri.ucsd.edu/Research/whatisfmri.html. Published 2022. Accessed 4 Aug 2022
6. Power JD, Cohen AL, Nelson SM et al (2011) Functional network organization of the human brain. Neuron 72(4):665–678. PMC3222858
7. O'Reilly JX, Woolrich MW, Behrens TE, Smith SM, Johansen-Berg H (2012) Tools of the trade: psychophysiological interactions and functional connectivity. Soc Cogn Affect Neurosci 7(5):604–609. PMC3375893
8. Lv H, Wang Z, Tong E et al (2018) Resting-state functional MRI: everything that nonexperts have always wanted to know. AJNR Am J Neuroradiol 39(8):1390–1399. PMC6051935
9. van den Heuvel MP, Hulshoff Pol HE (2010) Exploring the brain network: a review on resting-state fMRI functional connectivity. Eur Neuropsychopharmacol 20(8):519–534
10. Wang J, Zuo X, He Y (2010) Graph-based network analysis of resting-state functional MRI. Front Syst Neurosci 4:16. PMC2893007
11. Volkow ND, Koob GF, Croyle RT et al (2018) The conception of the ABCD study: from substance use to a broad NIH collaboration. Dev Cogn Neurosci 32:4–7. PMC5893417
12. Lisdahl KM, Tapert S, Sher KJ et al (2021) Substance use patterns in 9-10 year olds: baseline findings from the adolescent brain cognitive development (ABCD) study. Drug Alcohol Depend 227:108946
13. Barch DM, Burgess GC, Harms MP et al (2013) Function in the human connectome: task-fMRI and individual differences in behavior. NeuroImage 80:169–189. PMC4011498
14. Cohen N, Moyal N, Lichtenstein-Vidne L, Henik A (2016) Explicit vs. implicit emotional processing: the interaction between processing type and executive control. Cogn Emot 30(2):325–339
15. Knutson B, Westdorp A, Kaiser E, Hommer D (2000) FMRI visualization of brain activity during a monetary incentive delay task. NeuroImage 12(1):20–27
16. Logan GD (ed) (1994) On the ability to inhibit thought and action: a users' guide to the stop signal paradigm. Academic Press, Amsterdam
17. Casey BJ, Cannonier T, Conley MI et al (2018) The Adolescent Brain Cognitive Development (ABCD) study: imaging acquisition across

21 sites. Dev Cogn Neurosci 32:43–54. PMC5999559

18. Lejuez CW, Read JP, Kahler CW et al (2002) Evaluation of a behavioral measure of risk taking: the Balloon Analogue Risk Task (BART). J Exp Psychol Appl 8(2):75–84
19. Hulvershorn LA, Hummer TA, Fukunaga R et al (2015) Neural activation during risky decision-making in youth at high risk for substance use disorders. Psychiatry Res 233(2): 102–111. PMC4536117
20. Butcher TJ, Dzemidzic M, Harezlak J, Hulvershorn LA, Oberlin BG (2021) Brain responses during delay discounting in youth at high-risk for substance use disorders. Neuroimage Clin 32:102772. PMC8414537
21. Crane NA, Gorka SM, Weafer J, Langenecker SA, de Wit H, Phan KL (2017) Preliminary evidence for disrupted nucleus accumbens reactivity and connectivity to reward in binge drinkers. Alcohol Alcohol 52(6):647–654. PMC5860287
22. Carlson JM, Foti D, Mujica-Parodi LR, Harmon-Jones E, Hajcak G (2011) Ventral striatal and medial prefrontal BOLD activation is correlated with reward-related electrocortical activity: a combined ERP and fMRI study. NeuroImage 57(4):1608–1616
23. Cohen AO, Breiner K, Steinberg L et al (2016) When is an adolescent an adult? Assessing cognitive control in emotional and nonemotional contexts. Psychol Sci 27(4):549–562
24. Blair KS, Aloi J, Crum K et al (2019) Association of different types of childhood maltreatment with emotional responding and response control among youths. JAMA Netw Open 2(5):e194604. PMC6632148
25. Aloi J, Blair KS, Crum KI et al (2018) Adolescents show differential dysfunctions related to alcohol and cannabis use disorder severity in emotion and executive attention neuro-circuitries. Neuroimage Clin 19:782–792. PMC6031867
26. Makrushin A, Neubert T, Dittmann J (2017) Automatic generation and detection of visually faultless facial morphs. Proceedings of the 12th International Joint Conference on Computer Vision, Imaging and Computer Graphics Theory and Applications (VISIGRAPP) (6):39–50. https://doi.org/10.5220/0006131100390050
27. Windhager S, Bookstein FL, Mueller H, Zunner E, Kirchengast S, Schaefer K (2018) Calibrating facial morphs for use as stimuli in biological studies of social perception. Sci Rep 8(1):6698. PMC5923288
28. Roberts G, Green MJ, Breakspear M et al (2013) Reduced inferior frontal gyrus activation during response inhibition to emotional stimuli in youth at high risk of bipolar disorder. Biol Psychiatry 74(1):55–61
29. Hare TA, Tottenham N, Davidson MC, Glover GH, Casey BJ (2005) Contributions of amygdala and striatal activity in emotion regulation. Biol Psychiatry 57(6):624–632
30. Verbruggen F, Logan GD (2009) Models of response inhibition in the stop-signal and stop-change paradigms. Neurosci Biobehav Rev 33(5):647–661. PMC2696813
31. Lempert KM, Steinglass JE, Pinto A, Kable JW, Simpson HB (2019) Can delay discounting deliver on the promise of RDoC? Psychol Med 49(2):190–199
32. Balodis IM, Potenza MN (2015) Anticipatory reward processing in addicted populations: a focus on the monetary incentive delay task. Biol Psychiatry 77(5):434–444. PMC4315733
33. Aloi J, Blair KS, Crum KI et al (2020) Alcohol use disorder, but not cannabis use disorder, symptomatology in adolescents is associated with reduced differential responsiveness to reward versus punishment feedback during instrumental learning. Biol Psychiatry Cogn Neurosci Neuroimaging 5(6):610–618. PMC7286796
34. Schlund MW, Siegle GJ, Ladouceur CD et al (2010) Nothing to fear? Neural systems supporting avoidance behavior in healthy youths. NeuroImage 52(2):710–719
35. Ernst M, Nelson EE, Jazbec S et al (2005) Amygdala and nucleus accumbens in responses to receipt and omission of gains in adults and adolescents. NeuroImage 25(4):1279–1291
36. Rao U, Sidhartha T, Harker KR, Bidesi AS, Chen LA, Ernst M (2011) Relationship between adolescent risk preferences on a laboratory task and behavioral measures of risk-taking. J Adolesc Health 48(2):151–158. PMC3050646
37. Shad MU, Bidesi AS, Chen LA, Thomas BP, Ernst M, Rao U (2011) Neurobiology of decision-making in adolescents. Behav Brain Res 217(1):67–76. PMC3004983
38. Fox MD, Snyder AZ, Vincent JL, Corbetta M, Van Essen DC, Raichle ME (2005) The human brain is intrinsically organized into dynamic, anticorrelated functional networks. Proc Natl Acad Sci U S A 102(27):9673–9678. PMC1157105
39. Van Dijk KR, Hedden T, Venkataraman A, Evans KC, Lazar SW, Buckner RL (2010) Intrinsic functional connectivity as a tool for human connectomics: theory, properties, and optimization. J Neurophysiol 103(1): 297–321. PMC2807224

40. Bjork JM, Gilman JM (2014) The effects of acute alcohol administration on the human brain: insights from neuroimaging. Neuropharmacology 84:101–110. PMC3971012
41. Karoly HC, Schacht JP, Meredith LR et al (2019) Investigating a novel fMRI cannabis cue reactivity task in youth. Addict Behav 89: 20–28. PMC6309857
42. Squeglia LM, Jacobus J, Tapert SF (2009) The influence of substance use on adolescent brain development. Clin EEG Neurosci 40(1): 31–38. PMC2827693
43. Meruelo AD, Castro N, Cota CI, Tapert SF (2017) Cannabis and alcohol use, and the developing brain. Behav Brain Res 325 (Pt A):44–50. PMC5406224
44. Huettel SA, Song AW, McCarthy G (2014) Functional magnetic resonance imaging. Sinauer Associates, Oxford
45. James CJ, Hesse CW (2005) Independent component analysis for biomedical signals. Physiol Meas 26(1):R15–R39
46. Elliott ML, Knodt AR, Ireland D et al (2020) What is the test-retest reliability of common task-functional MRI measures? New empirical evidence and a meta-analysis. Psychol Sci 31(7):792–806. PMC7370246
47. Chen G, Pine DS, Brotman MA, Smith AR, Cox RW, Haller SP (2021) Trial and error: a hierarchical modeling approach to test-retest reliability. NeuroImage 245:118647
48. Haller SP, Chen G, Kitt ER et al (2022) Reliability of task-evoked neural activation during face-emotion paradigms: effects of scanner and psychological processes. Hum Brain Mapp 43(7):2109–2120. PMC8996353
49. West KL, Zuppichini MD, Turner MP et al (2019) BOLD hemodynamic response function changes significantly with healthy aging. NeuroImage 188:198–207. PMC6450381
50. Arichi T, Fagiolo G, Varela M et al (2012) Development of BOLD signal hemodynamic responses in the human brain. NeuroImage 63(2):663–673. PMC3459097
51. Pujol J, Blanco-Hinojo L, Macia D et al (2021) Differences between the child and adult brain in the local functional structure of the cerebral cortex. NeuroImage 237:118150
52. Guyer AE, Pérez-Edgar K, Crone EA (2018) Opportunities for neurodevelopmental plasticity from infancy through early adulthood. Child Dev 89(3):687–697. PMC5948168
53. Tooley UA, Bassett DS, Mackey AP (2021) Environmental influences on the pace of brain development. Nat Rev Neurosci 22(6): 372–384. PMC8081006
54. Giedd JN, Raznahan A, Alexander-Bloch A, Schmitt E, Gogtay N, Rapoport JL (2015) Child psychiatry branch of the National Institute of Mental Health longitudinal structural magnetic resonance imaging study of human brain development. Neuropsychopharmacology 40(1):43–49. PMC4262916
55. Mills KL, Goddings AL, Clasen LS, Giedd JN, Blakemore SJ (2014) The developmental mismatch in structural brain maturation during adolescence. Dev Neurosci 36(3-4):147–160
56. Sussman D, Leung RC, Chakravarty MM, Lerch JP, Taylor MJ (2016) The developing human brain: age-related changes in cortical, subcortical, and cerebellar anatomy. Brain Behav 6(4):e00457. PMC4802426
57. Mazziotta J, Toga A, Evans A et al (2001) A probabilistic atlas and reference system for the human brain: international consortium for brain mapping (ICBM). Philos Trans R Soc Lond Ser B Biol Sci 356(1412):1293–1322. PMC1088516
58. Joshi S, Davis B, Jomier M, Gerig G (2004) Unbiased diffeomorphic atlas construction for computational anatomy. NeuroImage 23 (Suppl 1):S151–S160
59. Sanchez CE, Richards JE, Almli CR (2012) Age-specific MRI templates for pediatric neuroimaging. Dev Neuropsychol 37(5): 379–399. PMC3399736

Chapter 6

Advanced PET Neuroimaging Techniques in Alcohol Use Disorder

Karmen K. Yoder

Abstract

When evaluating the alcohol research neuroimaging literature, it is vitally important to understand the fundamental aspects about the technologies and methods that can contribute to ambiguous results and/or conflicting data between studies. These nuances are not always obvious to non-imagers, or even neuroimagers who specialize in a single imaging modality. Here, we will attempt to address some key methodological considerations and potential pitfalls that can impact how the neuroligand positron emission tomography (PET) literature in AUD is interpreted. The discussions only consider research-grade PET neuroimaging acquisitions and not clinical diagnostic approaches.

Key words Alcohol use disorder, Positron emission tomography, Radioligands, Neuroimaging, Tracer kinetic modeling, Alcohol, Dopamine

1 Introduction

PET is a powerful in vivo molecular imaging technology that permits us to ascertain information about neurotransmitter systems that may inform us about the development and maintenance of AUD and how the brain responds to alcohol intoxication. Research neuroligand PET is a complex technique that requires extensive knowledge about data acquisition, data processing, and outcome metrics in order to critically assess and interpret the literature. This chapter is designed to provide basic concepts behind dynamic PET imaging, examples of image processing steps, the foundation of tracer kinetic modeling, quantitative parameters and semi-quantitative outcome metrics, and how one can use neuroligand PET to derive information about neurotransmitter release.

Melissa A. Cyders (ed.), *Translational Research Methods for Alcohol Use Disorders*, Neuromethods, vol. 201,
https://doi.org/10.1007/978-1-0716-3267-3_6,

2 Participant Characterization

Proper characterization of research participants is vital to understanding potential sources of variance in the data, regardless of the population sample being studied. The first step is selection of the target population that will best answer the hypothesis to be tested. Historically, most groups focused on individuals with severe AUD during acute detoxification and/or during abstinence. However, this may not be generalizable to the larger population of AUD, who continue to drink despite adverse consequences in their lives and who do not seek medical treatment for their drinking [1, 2]. This latter group may have a slightly less severe AUD profile and may have more subtle differences with respect to brain structure and function. There is also considerable interest in understanding the consequences of alcohol intake in young adults and adolescents; for example, binge drinking in these groups can confer greater risk for development of AUD.

Regardless of the group being studied, extensive characterization should be conducted. For example, we find it useful to determine current and prior medical conditions/medications (*see* **Note 1**), drinking frequency and lifetime use, mood states and traits, and aspects of personality that may influence drinking behaviors (i.e., impulsivity). Standard neuropsychological tests may also be germane for hypothesis testing in brain regions related to psychomotor and cognitive functions.

There is some debate as to how "clean" a participant sample should be. Funding agencies and reviewers often demand that studies on AUD exclude participants for comorbidities such as substance use disorders and/or psychiatric diagnoses. Scientifically, this is understandable, as we seek to identify the neural substrates of AUD in the absence of any confounding variables. However, we and our collaborators across the United States have observed that it is highly unusual to find participants that "only" have AUD. It is not uncommon for heavy-drinking participants and controls to test positive for drugs of abuse, including opiates, amphetamines, and, most frequently, marijuana (indeed, the legalization of medicinal and recreational marijuana in many states caused an uptick in reported usage across both otherwise healthy controls and participants with an AUD). Capturing information about frequency and duration of use of these substances is important for making decisions about whether or not to include subjects in the study, and whether to attempt to balance control groups for such usage. Another commonly reported comorbid diagnosis is affective disorders (e.g., depression, anxiety, bipolar disorder), which may or may not be treated with medication. Given that the neural substrates of these disorders remain unclear, the investigator must make an informed decision as to whether or not the presence of these diagnoses (and the severity of symptomatology) may confound the ability to detect effects believed to be specific to AUD (*see* **Note 2**).

3 Alcohol Exposure

An important area of inquiry in ascertaining the determinants of the development and maintenance of AUD is to understand the effects of acute alcohol intoxication on brain structure, function, and neurochemistry. Oral alcohol consumption has been used in many PET paradigms [3–7]. However, oral dosing is plagued by several biological factors that lead to high variability in blood alcohol concentration, and hence brain alcohol exposure, across individuals [8–10]. This variability in tissue exposure may result in unwanted confounds on primary outcome indices and inability to detect significant effects of acute intoxication. For example, age, sex, height, and weight all contribute to individual differences in rates of alcohol absorption, distribution, and elimination. In order to better control brain alcohol exposure across individuals, our colleagues developed an intravenous (IV) method of alcohol administration, the physiologically-based pharmacokinetically modeled (PBPK) IV alcohol protocol [8–10]. The PBPK controls the exact timing and concentration of alcohol delivery (*see* Sect. 9 on using PET to index neurotransmitter release for relevance of timing during imaging paradigms) and minimizes the experimental variation in the brain exposure to alcohol across subjects. The rate and concentration of alcohol infusion during the IV procedure are pre-determined by the 3-compartment PBPK model, which takes into account the rates of alcohol absorption, distribution, and elimination based on the age, sex, height, and weight of the participant. This model can be used to precisely control the rate, "dose," and duration of alcohol exposure and can also be adapted for controlled self-administration paradigms [11, 12]. Although there is some debate as to the ethnological relevance of the IV paradigm compared to oral drinking behaviors [13], many neuroimaging laboratories choose to use the PBPK IV protocol in order to achieve precision and accuracy in brain alcohol exposure across individuals [14–22].

4 Positron Emission Tomography (PET)

The underlying principle of PET is the utilization of a pharmacologically insignificant amount of radioactive substance to observe physiological systems within the body. The radioactive "tracer" is labeled with a positron-emitting radionuclide and is injected intravenously (there are other delivery methods, but IV is the most common). Shortly after the positron (a.k.a., e + or β; same mass as an electron, but with a positive charge) is emitted from the tracer molecule, it almost immediately collides ("annihilates") with an electron in the immediate environment. The product of this annihilation event is two gamma particles (photons) that are emitted at 180° from each other (*see* Fig. 1). The PET scanner detects the detector location, number, and timing of the photons emitted from the body ("events"). Reconstruction algorithms take information about the detected events, and from this information produce reasonably accurate estimates of where in the body the annihilation event occurred (*see* **Note 3**). The end result is a series of images over time (*see* Sect. 5) that contain information about the amount of radioactivity in the tissue at each voxel in the image (Bq/cc).

The most common diagnostic application of PET for brain disorders is [^{18}F]flurodexoyglucose (FDG) for assessing glucose uptake in the brain as a proxy for neuronal activity. For research hypotheses, FDG has largely been replaced by magnetic resonance imaging (MRI) techniques that index relative brain activity with superior temporal and spatial resolution (functional MRI) or which can quantitate blood flow (i.e., arterial spin labeling). FDG is commercially available and FDG scans could be acquired at most major medical centers. However, most research PET tracers used to study the neurochemistry of the brain require an on-site cyclotron

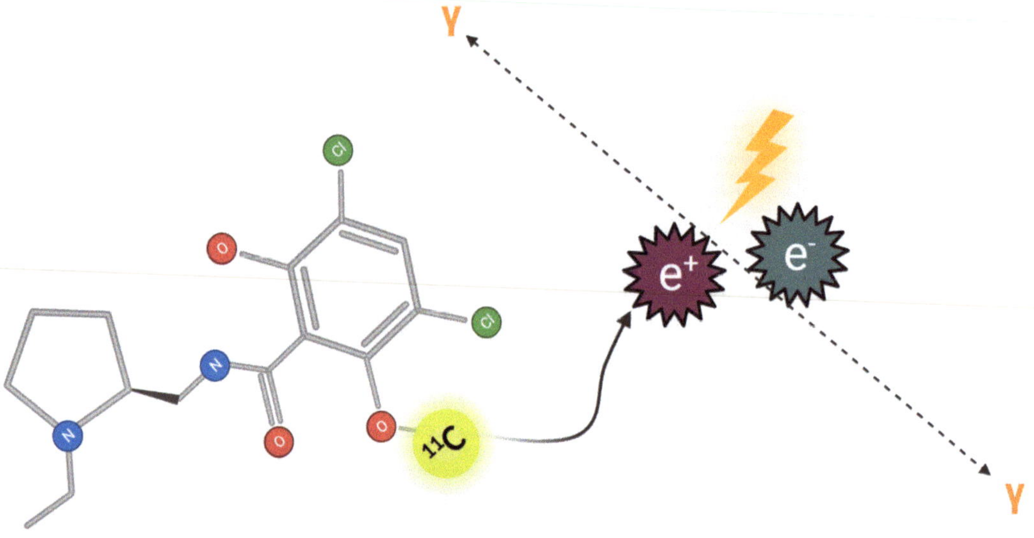

Fig. 1 Positron emission. Raclopride, a D_2/D_3 receptor antagonist, can be labeled with the positron-emitting radionuclide, [^{11}C] (yellow). When radioactive decay occurs, a positron (e^+) is emitted from the nucleus, and almost immediately encounters an electron (e^-) in the environment. The two particles, which are of the same mass but opposite charge, "annihilate" one another. The product of this annihilation event is two gamma particles (γ; photons) which are emitted at a 180° angle from one another. These photons are what are detected by the PET scanner. Created with BioRender.com. [^{11}C]raclopride chemical schematic adapted from https:/pubchem.ncbi.nlm.nih.gov/compound/Raclopride-C-11#section=2D-Structure

and radiochemistry facility, which restricts their use to roughly a dozen sites across the USA. The most commonly studied neurochemical system in AUD is dopamine (DA), in particular, with the DA D_2/D_3 receptor tracers [^{11}C]raclopride and [^{18}F]fallypride, which have different properties, and thus lend themselves to different types of hypothesis testing and study designs [23]. Tracers also exist for D_1 and D_3 receptors, DA and norepinephrine transporters, serotonergic receptors, metabotropic glutamate receptors, nicotinic receptors, opiate receptors, benzodiazepine binding sites, and other esoteric targets that may relate to the etiology of AUD. Collectively, these ligands are less frequently used in the study of AUD for various logistic and analytic complications and because of complex site-specific radiosynthesis protocols.

5 Dynamic and Static Imaging

Regardless of the tracer to be used, there are two basic types of PET acquisition protocols: dynamic imaging and "static" imaging. With dynamic imaging, the goal is to observe the continuously changing behavior of the tracer over time to derive physiologically-relevant parameters about the target of interest. This temporally sensitive information permits quantitative data analysis through tracer kinetic modeling [24, 25] (*see* Sect. 7 on quantitative metrics). Image acquisition begins immediately upon injection of the tracer (tracer administration could be either a single bolus infusion, or a bolus + continuous infusion, in which a desired level of radioactivity is maintained throughout the scan). The data (the number, timing, and detector location of events) are acquired continuously for a pre-specified time period that is dependent on a combination of factors such as radionuclide half-life, kinetics of the tracer, and, in some cases, a priori hypotheses about the brain region of interest. Event data are binned into short time "frames," which often increase in length over time to maintain counting statistics of events as the radioactivity in the body decays (although some laboratory protocols may use only short bins due to post-processing preferences and/or analytical needs). A reconstructed image is generated for each time frame; the intensity value at each voxel is the average Bq/cc for that time frame, adjusted for decay correction and weighted for the length of the frame. With dynamic imaging, the temporal profile of the tracer will vary widely, depending on the kinetics of the tracer (*see* Fig. 2).

"Static" imaging refers to image acquisition that takes place after a tracer uptake period. This type of protocol yields semi-quantitative data. Tracer is injected with a single bolus infusion, and the participant usually waits in a comfortable chair for a given amount of time (this could range from 30 min to over an hour, again, depending on the radionuclide and kinetics of the tracer).

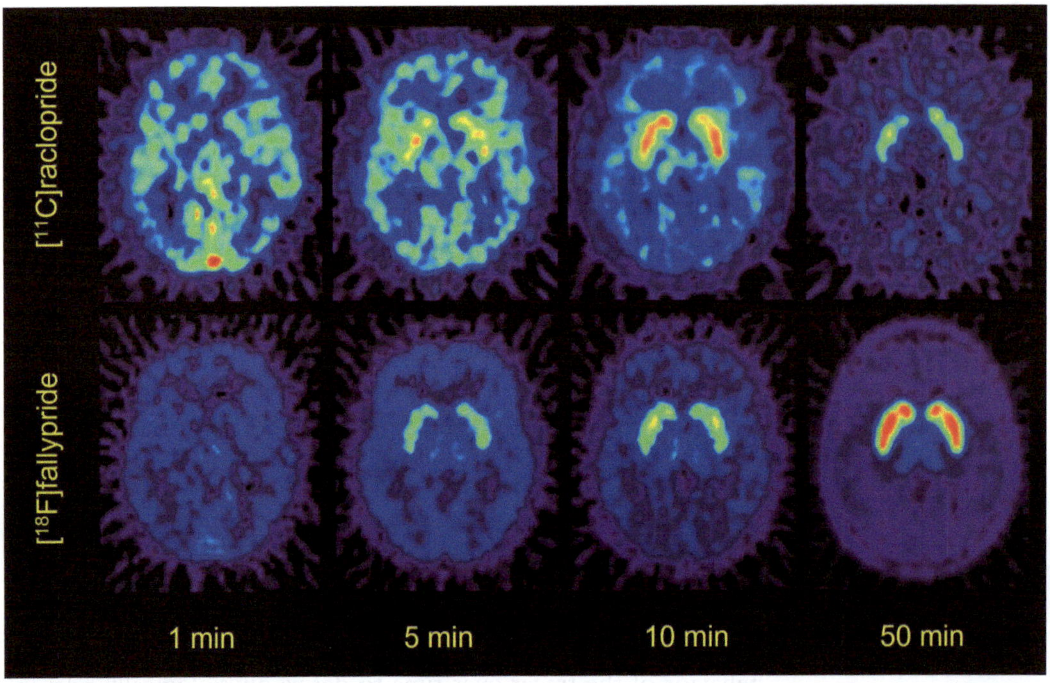

Fig. 2 Tracer kinetics vary by radioligand. Both [^{11}C]raclopride (top) and [^{18}]fallypride (bottom) are dopamine D_2/D_3 antagonists. Axial images that feature the striatum are shown at *x* min after IV tracer injection. [^{11}C]raclopride has relatively fast kinetics and has started to "wash out" of the brain by 50 min. [^{18}]fallypride has a slow off-rate and still has high specific binding at 50 min

Typically, a set of 5 or 6 five-minute frames are acquired. The assumption with most tracers acquired in this manner is that the tracer has come to equilibrium and the tracer behavior is stable (unchanging, or "static"). In this case, the acquisition of multiple frames is not meant for observing change over time; instead, the average radioactivity in the tissue is averaged across the frames to decrease statistical noise. Typically, this is done by creating a single average image during post-processing.

For both dynamic and static protocols, it is standard practice to also acquire a structural T1-weighted magnetic resonance image (MRI) for processing of the PET emission data. Most research laboratories are currently using 3 T MRI scanners.

6 PET Image Post-processing and Quality Control

Modern PET scanners create four-dimensional image output in the standard DICOM format. In our experience, the majority of software platforms used by neuroligand PET researchers cannot handle DICOM data, and instead read in the Neuroimaging Informatics Technology Initiative (NIfTI; https://nifti.nimh.nih.gov/) format. NIfTI has the advantage of preserving left/right information, but temporal information about the frame times is lost. Although many platforms still recognize the older Analyze format, the left/right information is not always preserved correctly or consistently and should be avoided (*see* **Note 4**). Many software platforms offer DICOM to NIfTI conversion functions (*see* **Note 5**).

All PET scanners utilize a short acquisition of either a "transmission" scan, or a low-resolution computed tomography (CT) scan to perform attenuation correction (AC) on the PET data. Both types of scans send a form of gamma radiation through the head to measure the degree to which gamma rays are attenuated by the skull; a correction factor is derived and applied to the PET data to account for the photons emitted from the brain that were attenuated by the skull. The AC scans are acquired separately from the PET data; the assumption is that the participant remains in the exact same bed position for the AC and PET scan so that the brain position in the AC and PET space are identical. Although the AC scan is typically acquired only a few minutes before the PET data are acquired, there is still a chance that the subject may move between the scans. If the motion is severe, then the PET data will not be reconstructed correctly. This often results in an apparent lateralization of signal, such that it seems that the radioactivity is higher in one part of the brain. Undetected, this would contribute unwanted variance to the dataset. Thus, the first QC step must be to visually assess whether the AC scan lines up with the PET emission data. Once there is confirmation that the AC scan and emission data are in the same space, additional post-processing may continue (*see* **Note 6**).

Both dynamic and static acquisitions are subject to participant motion, which can create artifacts within the image dataset. True motion correction requires fiducial markers and application of sophisticated algorithms at stages prior to image reconstruction [26]; this software is not commercially available and is not part of the infrastructure at even the most advanced research sites. Instead, we advocate for rigorous quality control procedures to identify motion and simple steps to attempt to correct for it to the extent possible with available algorithms.

Visual inspection of sequential time frames can help identify gross motion. This could be done either by visualizing the frames side by side (e.g., "Check Reg" function in Statistical Parametric Mapping (SPM) software) or via a cine loop. However, some motion is subtle and may not be easily identified visually. Thus, it is important to determine motion empirically. Our laboratory does this in two ways along our processing stream. The general steps are as follows. (Although we utilize SPM (https://www.fil.ion.ucl.ac.uk/spm/), other free programs that offer comparable algorithms are available, such as Analysis of Functional NeuroImages (AFNI; https://afni.nimh.nih.gov/) and FMRIB Software Library (FSL; https://fsl.fmrib.ox.ac.uk/fsl/fslwiki/).

6.1 MRI Post-processing

- Convert structural T1 MRI DICOM data to NIfTI.
- Coregister participant's T1 to, e.g., the SPM avg152T1 template.
- Run a segmentation algorithm on the participant's T1 that has been coregistered to the template. This will generate probability estimate maps for gray matter, white matter, and cerebrospinal fluid.
- If a voxel-wise analysis is planned, generate a transformation matrix or deformation field for the resliced T1 image. This will allow you to spatially warp images from the resliced T1 space to, e.g., Montreal Neurological Institute (MNI) template space.

6.2 PET Post-processing

- Convert PET DICOM data to NIfTI (*see* **Note 5**).
- Visually check the native PET frames. Identify the first frame with a faint outline of the brain.
- Determine initial degree of motion in [*x*, *y*, *z*] and pitch, yaw, roll. (In SPM12, this is the "realign: estimate" function). Read in all images. Save the output of the motion parameters for future reference.
- Create a mean PET image that contains a mix of blood flow and specific binding (*see* **Note 7**). Record the frames used to create the mean PET.
- Visually examine mean PET to determine image quality and suitability for registration with early and late time frames.
- Coregister the mean PET to the resliced T1.
- Visually examine the coregistration between the resliced mean PET and the resliced T1.
- Coregister all emission data (starting with the first frame with a faint brain outline) to the resliced mean PET. This will bring all images into resliced T1 space.
- Rerun the motion parameter estimates on the resliced emission data to determine how much residual motion remains. Save the output of the motion parameters for comparison to the initial estimate and for future reference.
- Extract time-activity curves (TACs; tissue radioactivity, Bq/cc, over time) from at least one a priori region of interest that contains specific binding (left and right), and plot to inspect for noise in the TACs that may indicate motion and/or problematic frames that need to be removed (*see* **Note 8**).
- If voxel-wise analyses are to be conducted, apply the appropriate deformation field/transformation matrix from the resliced T1 to spatially warp the emission images (which are in resliced T1 space) into MNI space.

- We recommend that region of interest analyses be conducted in the resliced T1 image space, and not MNI space (*see* **Note 9**). It is often easiest to start with automated extractions of anatomically defined regions (e.g., generated by Freesurfer or similar programs), and edit them relative to each participant's resliced T1 for anatomical precision.

7 Quantitative Metrics: V_D and BP

For simplicity, we will discuss quantitation with respect to reversible, competitive ligands. The goal is to obtain quantitative information about the relative receptor "availability" within the image. We don't actually know "where" a tracer is within a voxel, because it could be in any one of several states (*see* Fig. 3). To derive meaningful physiological information about our target, we employ tracer kinetic models. In brief, these are compartmental models that describe the different theoretical (or physical) states that a tracer could be in within the tissue (*see* Fig. 4). There are sets of differential equations that describe the movement of the tracer between these states; solutions to these equations yield rate constants that are related to receptor density (although *see* Sect. 9 for contributions of endogenous ligand to the signal). In order to generate solutions to the differential equations in the tracer kinetic model, (that is, to get the best estimate of the relative degree of specific binding of the tracer), we have to provide the model with a known input function that provides information of how much radioactivity the tissue is receiving from the arterial plasma over time. Under particular circumstances, we may instead provide the model with TACs from voxels or regions within the brain or from vascular areas. With the "known" information of tracer behavior in the plasma, we can then solve the differential equations to describe the movement and states of the tracer in the brain. Details about the derivation and solutions of these models are described elsewhere [24, 25].

The canonical gold standard for the input function is the metabolite-corrected TAC of the parent radioligand in the arterial plasma. However, arterial input functions are difficult to obtain and can be very painful in some populations; the sample processing and analyses require a lot of infrastructure, and the data can also be plagued by noise from sampling or processing errors. Thus, many sites prefer to use alternative input functions. These most often are "reference region" approaches, which utilize TACs from a brain region devoid of specific binding of the radioligand. Image-derived input functions from large arteries may also be acceptable in some cases. For any given tracer, alternative input functions must always be validated against metabolite-corrected arterial input methods (*see* **Note 10**).

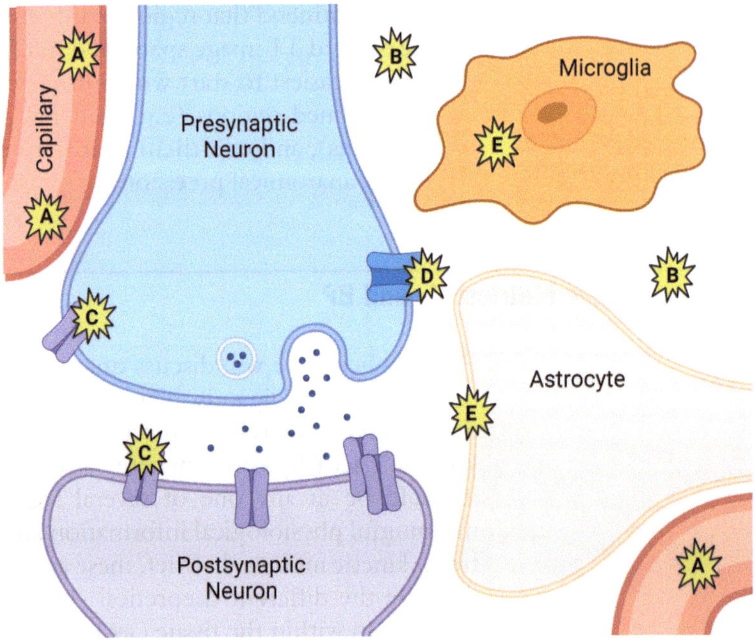

Fig. 3 Potential plasma and tissue states of a radiotracer (yellow stars) within a PET image voxel. The tracer can exist in multiple states, denoted by letters. (**a**) Within the plasma. (**b**) Within the interstitial tissue. (**c**) Specifically bound to the target receptor, which may exist both pre- and post-synaptically. (**d**) Non-selective specific binding to another target. (**e**) Non-specifically bound to tissue either within or on a cell. PET images provide the concentration of radioactivity within the tissue (Bq/cc), which is a summation of all the tracer states. In order to get to estimates of specific binding, tracer kinetic modeling is needed. Created with BioRender.com

An extensive comparison of all the potential model-based and linearized approaches to obtain quantitative estimates of relevant parameters is beyond the scope of this chapter. Instead, we will focus on the two primary quantitative endpoints of interest, volume of distribution (V_D) and binding potential (BP). V_D describes the ratio of the concentration of radioactivity in the tissue relative to the concentration of radioactivity in the blood at equilibrium. That is, if $V_D = 2$, then the interpretation is that the tissue concentration of radiotracer is twice that of the [radiotracer] in plasma. Said another way, 2 mL of blood has the same [radiotracer] as 1 mL of tissue. At the simplest level, it is operationally defined as the ratio of the influx and efflux constants, K_1 and k_2 (*see* Fig. 4). Note that, on the surface, this metric does not provide any direct information about the location of the tracer in the tissue voxel or region. However, it does provide a reasonable description of the degree of tracer retention within the tissue. Binding potential, more commonly referred to as "receptor availability", provides an estimate of receptors available for binding, and is operationally defined as the ratio of

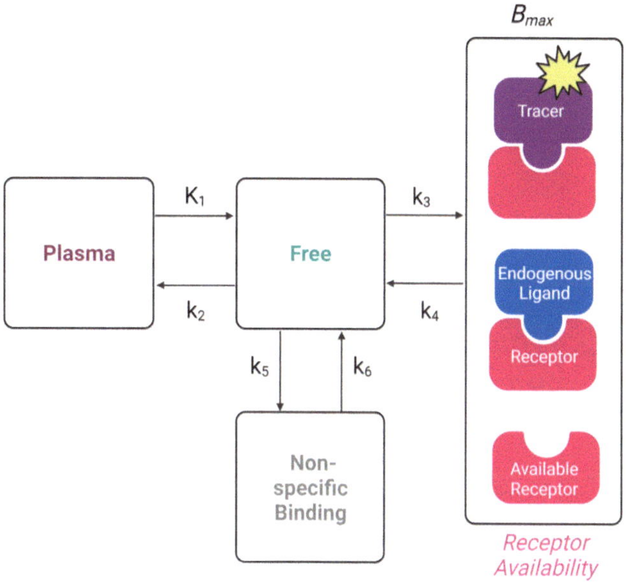

Fig. 4 An example of a tracer kinetic compartmental model. The compartments describe theoretical states in which a tracer may exist within brain tissue. This is a "three-tissue" compartment model: the tracer may exist as a free (unbound) molecule (middle), as specifically bound (far right box), or as non-specifically bound (lower box). Plasma is considered an input function to the model and is not a tissue compartment. The arrows denote potential movement of the tracer between compartments. These movements are described by differential equations. The solutions to these differential equations yield rate constants (K, k) that provide estimations of quantitative parameters that describe tracer retention ($V_D \equiv K_1/k_2$) and receptor availability ($BP \equiv k_3/k_4$). Although the non-specific compartment is shown here for descriptive purposes, in practice, it is often not utilized. B_{max}, maximal receptor density. Created with BioRender.com

the on- and off-rates of the radioligand to the target receptor (*see* Fig. 4; k_3/k_4). The on-rate constant, k_3, is a compound constant of two Michaelis-Menten kinetic parameters, $k_{on} * B_{max}$. Similarly, the off-rate constant, k_4, is equivalent to k_{off}. The binding affinity of the tracer, $K_D = k_{off}/k_{on}$. Thus, BP can also be defined in terms of Michaelis-Menten kinetics as B_{max}/K_D (maximal binding density relative to the binding affinity of the tracer). Both V_D and BP have various mathematical definitions based on the model or linearization method employed for parameter estimation [27]. Regardless of formulation, these are the primary quantitative metrics that will be encountered in the literature.

Implementation of the quantitative parameter estimation requires that the model structure has been correctly implemented into a computationally favorable software format. At this writing, the commercially available packages can be prohibitively expensive. Most institutions that utilize quantitative approaches rely on "in-house" code written by individuals with expertise in tracer kinetic modeling.

8 Semi-quantitative Metrics: SUV, SUVR, and SUVR-1

There are several "semi-quantitative metrics" that are frequently utilized because they are simple and fast to implement, and do not require extensive computational infrastructure and tracer modeling expertise. Standardized uptake value (SUV) takes the radioactivity concentration from the image voxels and divides them by the injected dose of tracer per body weight of the individual. While the intent is to "normalize" the effect of total radioactivity injected (which will differ across individuals), SUV is problematic, and results with this metric should be interpreted with caution [28] (*see* **Note 11**). Depending on the tracer, SUV may not be a reliable substitute for quantitative indices such as BP (*see* Fig. 5, [28]).

Standardized uptake value ratio (SUVR) refers to image intensity normalization of voxels from a target region by the radioactivity concentration from a brain tissue that has little or no specific binding (analogous to the "reference region" approaches described above). For some tracers, SUVR may be acceptable as a proxy for quantitative parameter estimates, as shown in Fig. 5 for [^{11}C] raclopride, and as was demonstrated previously for [^{11}C]PIB [29]. More recently, a variation on SUVR, SUVR-1, has emerged as another viable proxy for BP_{ND} (specific binding relative to non-displaceable uptake, [30]), in this case, for the tracer [^{11}C]UCB-J, with the centrum semiovale as the reference tissue [30, 31]. The mathematical derivation of how SUVR-1 reasonably approximates BP_{ND} is beyond the scope of this chapter; however, it follows from established tracer kinetic modeling frameworks [25, 32].

The overall take-home message for all semi-quantitative metrics is that, like alternative input functions for quantitative methods, they must be validated against established canonical quantitative parameters, such as V_D and/or BP_{ND}. If a quantitative reference region approach has been validated for a tracer, comparing SUV, SUVR, and SUVR-1 to, e.g., BP_{ND} is acceptable.

Semi-quantitative metrics are simple calculations and do not require complex computational processes. Voxel-wise calculations can be performed within most neuroimaging processing platforms, and average values extracted from brain regions are easily handled with simple division functions.

9 Indexing Relative Changes in Neurotransmitter Levels

In Fig. 4, there are two interesting features in the right-hand schematic that describes potential states of the total target receptor pool: receptors bound by endogenous ligand, and unbound receptors. In a system in which both the radioligand and the endogenous ligand are competitive and reversibly bound, this presents an enormous

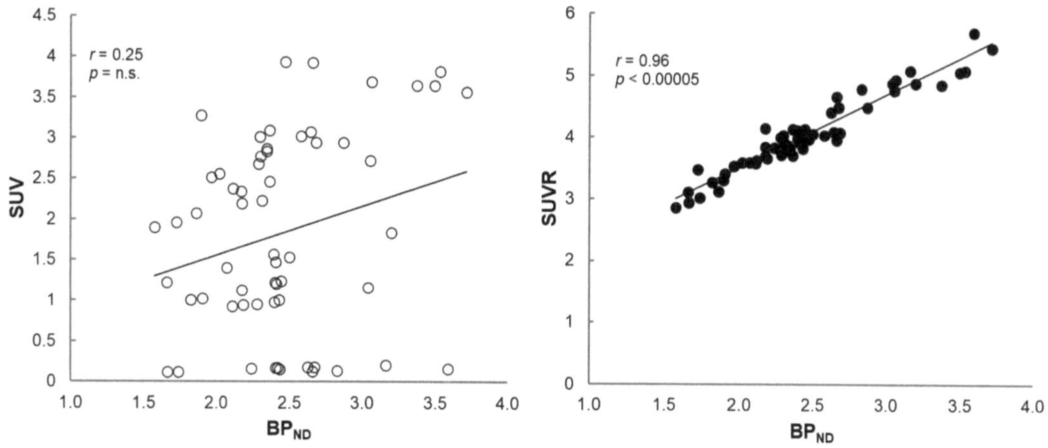

Fig. 5 Example of behaviors of semi-quantitative metrics SUV and SUVR relative to the quantitative index of BP_{ND} from striatal [^{11}C]raclopride data. Left: SUV (standardized uptake value) does not correlate with the quantitative metric BP_{ND}, (i.e., BP relative to nondisplaceable uptake, [27]). Right: Target region radioactivity normalized by radioactivity from a reference region (cerebellum), a.k.a. "standardized uptake value ratio" (SUVR). In this case, SUVR correlates well with BP_{ND}. Data were taken from twenty human subjects scanned with ~20 mCi [^{11}C]raclopride on a Siemens HR+. Data were generated from six regions of interest per subject: left and right post-commissural putamen, pre-commissural dorsal caudate, and ventral striatum [57]. Cerebellum was used as a reference region. BP_{ND} values were estimated with MRTM [58]

opportunity. We can exploit such a system to look at relative changes in neurotransmitter [32, 33]. To provide a concrete context, we will discuss this in terms of the endogenous neurotransmitter, dopamine (DA), and [^{11}C]raclopride (RAC), a competitive dopamine D_2/D_3 antagonist with favorable kinetics for displacement [34]. Put simply, if some experimental manipulation increases synaptic and extrasynaptic levels of dopamine, then DA will displace RAC and the overall PET signal will decrease relative to a baseline state. Conversely, if an experimental manipulation causes a decrease in DA levels, then there will be an increase in RAC signal as receptors become available for tracer binding. Several laboratories have taken advantage of this general scheme to investigate dopaminergic neurotransmission in AUD and how dopamine responds to alcohol intoxication [3–6, 20–22, 35, 36].

There are two general within-subjects methods to study changes in dopamine neurotransmission in response to an experimental manipulation: a two-scan approach in the case of single bolus injection studies, and a single-scan approach in the case of bolus-infusion studies (in which tracer concentration is maintained throughout the scan). For both designs, two components are key: a baseline (BL) acquisition, and precise (consistent) timing of the experimental manipulation (EXP). Figs. 6 and 7 illustrate the general scheme of both approaches and the time-activity curves that would be expected from each design. In the case of RAC, estimations of BP_{ND} are obtained from the baseline state and from the data acquired during the experimental manipulation. The % change

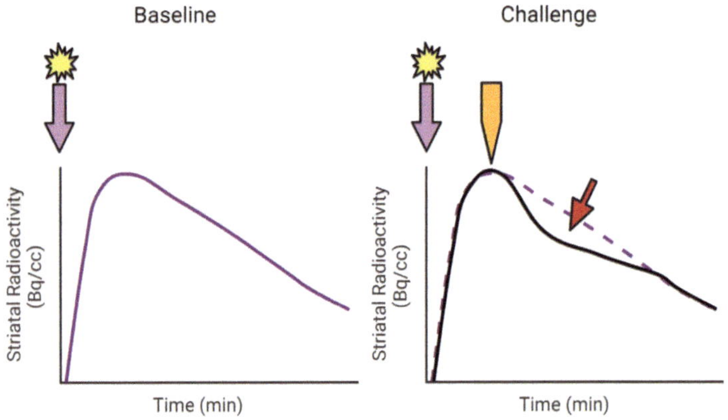

Fig. 6 Two-scan, single bolus approach for indexing changes in neurotransmitter release. Left: Baseline scan. Right: Experimental manipulation or "challenge" scan. Purple arrows with yellow stars denote a bolus tracer injection. The orange arrow indicates the beginning of the challenge, which could be pharmacological, cognitive, or motor. The red arrow indicates the deflection in the time-activity curve, which presumably is the result of increased neurotransmitter concentration that is displacing the radiotracer. For each scan, binding potential is estimated with the full time-activity curves. The deflection in the challenge curve will result in lower values of BP. Created with BioRender.com

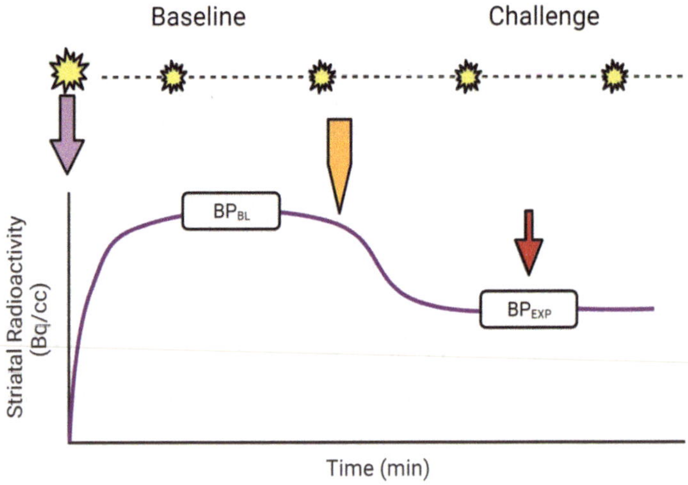

Fig. 7 Single scan, bolus + infusion approach for indexing changes in neurotransmitter release. The purple arrow with the yellow star denotes the initial bolus tracer injection; the dotted black line with yellow stars indicates that tracer is infused during the entire scan acquisition. The orange arrow indicates the beginning of the challenge, which could be pharmacological, cognitive, or motor. The red arrow indicates the deflection in the time-activity curve, which presumably is the result of increased neurotransmitter concentration that is displacing the radiotracer. In this paradigm, binding potential scan be estimated for both baseline (BP_{BL}) and challenge (BP_{EXP}) portions of the scan with a small amount of data at steady state (light yellow bars). Created with BioRender.com

in BP_{ND} ($\Delta BP = [(BP_{BL} - BP_{EXP})/BP_{BL}]$ is related to the relative changes in dopamine levels induced by the experimental condition. It is important to note that ΔBP is not strictly proportional to changes in DA levels [37], so no assertions about absolute change in DA concentration may be made with these methods (although *see* Sect. 10). The degree of observed ΔBP is a complex function of the timing, shape, and magnitude of the dopamine function [37]. Even if we assume that dopamine release (or decay) functions are similar across subjects, it remains imperative to keep the timing of the stimulus relative to tracer injection consistent for all participants to avoid unwanted experimental variance. The reality is that dopamine release (or decay) functions are highly likely to be inconsistent across individuals [21, 38], making experimental control even more important.

The basic types of experimental manipulations are pharmacological, cognitive, or motor. Most classical pharmacological studies include a placebo condition. Similarly, cognitive and motor studies often include a control condition that attempts to isolate potentially confounding effects of sensory or motor activation from the effect of interest (e.g., visual system activation during a cue reactivity study; visual and motor activation during a behavioral response study). However, our laboratory and others recommend a "true" resting baseline. Several RAC studies have shown that placebo itself increases DA levels [39–41], which, if used as a baseline condition, would reduce the ability to detect effects of increased DA from the experimental manipulation due to a potentially restrictive range of, e.g., BP values between conditions. Similarly, we have observed that small differences in instructions to participants results in significantly different dopaminergic states at "baseline" [42], again underscoring the need for precise experimental control of the cognitive state of participants (which, admittedly, is easier said than done). Unfortunately, radioligand PET is very costly, and it is not always possible to conduct the desired experimental controls of comparing a negative control state relative to a resting baseline. Investigators should acknowledge these possibilities, and exercise their scientific judgment as to which type of confounds they are willing to accept with respect to the ability to detect effects.

Finally, all tracers have some degree of within-subjects test-retest variance that must be considered within the context of detecting changes in neurotransmitter with radioligand PET. Speaking broadly, the field accepts test-retest reproducibility values of ±10% as being a reasonable upper limit. It is also possible to detect statistically significant effects of ΔBP that are below ±10% change. This sets up the debate as to whether or not statistically significant effects below the ±10% threshold are physiologically relevant. We will not endeavor to answer this question; rather, we will leave this point for the reader to consider within the context of their own work.

10 A Brief Word About Novel Approaches to Describe Neurotransmitter Dynamics

The metric ΔBP is coarse, in that it only provides information about whether, e.g., dopamine went "up" or "down". Potentially physiologically relevant (and exciting!) information about the dopamine system in AUD may reside in knowledge about the dynamics of dopamine release (or decay) functions: the timing of the peak of increased levels, the relative magnitude of increase, and the overall shape and/or exponential decay of the function. The possibility of deriving this information from neuroligand PET data was first explicitly recognized in the 1990s [32, 33] and has been further explored fairly extensively (for review, *see* [43]). Several methods for estimating dopamine curve parameters have been proposed and refined over the years [33, 36, 43–45]. While a thorough explication of this body of work is beyond the scope of this chapter, we wanted the reader to be aware that the field of neuroligand PET is still innovating and evolving to provide more specific information about dopaminergic neurotransmission that may yet yield novel information relevant to the development and maintenance of AUD. The field awaits development of suitable tracers for other receptor systems that would yield information about how the dynamics of, e.g., other monoamines, peptides, and amino acids contribute to AUD.

11 Notes

1. Humans can be surprisingly inaccurate and inconsistent when it comes to self-report questionnaires. Study personnel should closely supervise or directly administer any self-report indices. When ambiguous answers are provided, they should query for more specific information. Personnel should also monitor for discrepancies across instruments that may contain redundant information and obtain clarification. When possible, the use of very specific questions is preferred over more general items. For example, in our experience, "Do you have any major medical disorders" typically elicits a "no" on a basic phone pre-screen, which upon a more detailed in-person interview, may prove to be wildly incorrect. Diseases that were not endorsed as a "major medical disorder" on a pre-screen have included AIDS, diabetes, epilepsy, hepatitis C, and cancer, all of which disqualified the respective participants. Therefore, we recommend a more detailed inquiry that includes an extensive list of medical conditions ("Have you ever had [disease/disorder]"), even if it prolongs the pre-screen or in-person interview.

2. One factor that is unavoidable in studying AUD is the high comorbidity of cigarette smoking. Many studies do not include information about smoking behaviors in their samples. However, a growing body of evidence suggests that use of combustible smoke products such as cigarettes exerts a significant amount of variance on structural and functional measurements [46–51], even above and beyond heavy drinking [46, 48, 49, 51]. Thus, it is important to attempt to balance groups for smoking status if the primary factor of interest is the presence of an AUD. If the groups are unbalanced for smoking, it is possible that any effects detected in the AUD group may actually be mediated by smoking. Although currently there is no direct evidence that marijuana smoking has similar effects, it should be taken into consideration that cigarette and marijuana smoke share thousands of compounds that may have toxic consequences [52–56].
3. There are many different types of reconstruction algorithms, and they do not produce equivalent estimates of radioactivity within the brain. The two primary categories are analytical and iterative; discussion of these in any detail is beyond the scope of this chapter. However, when comparing studies that may have discrepant results, it is important to consider what algorithms were used in the reconstruction. They can produce very different ranges of Bq/cc within the image, and can have very different spatial and noise properties. These factors impact both quantitative and semi-quantitative metrics.
4. Despite all our advances in technology, differences in left/right display conventions still haunt researchers and can result in disastrous processing errors, especially if there are discrepancies between how MRI and PET data are handled. There are two conventions: radiological (left is displayed on the right side of the visual image, and vice versa), and neurological (left is displayed on the left, right is on the right). In general, neuroimaging researchers prefer neurological, whereas radiological still is standard for clinical purposes. One should confirm which convention your software uses for a default display, and whether or not a toggle between the two conventions is available. If possible, it is helpful to have a dataset or phantom in which there is no doubt which side is which (for example, presence of a clinically documented brain injury or tumor, or a phantom that has been digitally labeled). It is always best never to assume how different platforms handle image data.
5. DICOM to NIfTI converters do not provide an option to rename image files, which typically carry a long numerical designation assigned by the scanner. It would be worth creating a script that can rename all image frames with a unique and easily identifiable study ID to reduce potential input mistakes during processing and data analyses.

6. There is no pre-defined standard for how much spatial offset is acceptable between the AC scan and the emission scans. Our group has adopted the limit of the practical (not intrinsic) resolution of the PET scanner. For example, the practical resolution of the Siemens mCT is approximately 5 mM^3. Thus, if there is more than a 5 mM offset in any direction, we assume that the attenuation correction is corrupted and therefore the PET emission data cannot be used. An additional observation is that attenuation correction sometimes is readily apparent in the emission images as a visual artifact of signal laterality or other spatial imbalance of signal intensity.
7. The mutual information algorithms that are often used for coregistration and realignment functions depend upon, well, mutual information between the images. We have found that it is difficult to register most emission data directly to the T1 image: early emission frames do not have sufficient anatomic definition, whereas late frames may emphasize certain subcortical structures and contain little cortical definition (i.e., dopaminergic ligands that emphasize the striatum). Therefore, we create a "half-way" image that contains a mix of blood flow and specific binding. This "mean image" provides sufficient information to which all emission images may be successfully coregistered. How many frames are selected is completely ligand-dependent. Knowledge of how the tracer behaves over time and the timing of the acquisition protocol is required to determine this. It will be necessary to understand when the ligand typically begins to demonstrate specific binding (e.g., *see* Fig. 2).
8. There is no standard threshold for determining how much noise or movement is acceptable or unacceptable. Often, eliminating a few time frames in the middle or end of the scan is unlikely to significantly impact canonical quantitative estimates such as volume of distribution and/or binding potential. However, excessive noise and motion can confound the outcome variables. It is the responsibility of the investigator to determine whether or not a participant's data should be retained within a sample. Published characterizations of ligand stability (e.g., test-retest variability) and initial tracer kinetic modeling studies (including simulations) could be used as a guide. Investigators can also do internal testing on how data removal impacts quantitative estimates. Although the sheer cost of PET makes any data loss an unappetizing prospect, it is often necessary to maintain scientific integrity.
9. In our experience, defining regions of interest in MNI space and applying them to the spatially normalized emission data is not optimal. We have found that, even when warped to MNI

space, the normalized image anatomy of, e.g., participants with AUD who smoke is often very different from the template MNI image, especially with respect to the lateral ventricles and striatal substructures. Thus, we elect to conduct region of interest analyses in the resliced T1 space. Anatomical offset from the template also warrants caution when interpreting voxel-wise results.

10. Alternative input function methods must always be thoroughly characterized against the gold standard, the metabolite-corrected plasma arterial input function. This must be done for each radioligand individually. For example, let us say that tracers A and B are both DA D_2/D_3 antagonists. It has been shown that a reference region approach was validated against modeling with an arterial input function for Tracer A, but similar validation is not yet published for Tracer B. One cannot assume that a reference region approach will also be acceptable for Tracer B. To use a reference region approach (or e.g., an image-derived input function) for Tracer B without published validation is not best practice.

11. The metric SUV has several underlying assumptions. First, and most important, is that the tracer is equally distributed across the body, that is, all tissues have the same probability of being exposed to the tracer; a related assumption is that the tracer behaves similarly across individuals with respect to absorption, distribution, and elimination. These assumptions are almost automatically violated given the great variability of body weight and body composition across individuals, which result in differential tracer distribution between participants. Interestingly, we have observed that use of total body weight versus estimates of lean body weight result in very different values of SUV. Another assumption is that the tracer comes to some steady state that does not vary much over time; this is reasonably easy to verify from dynamic imaging studies, and is generally not a significant concern. However, SUV should not be used if tracers are not at apparent steady state during image acquisition. Finally, there is an assumption that the effects of any radioactive metabolites are negligible (that is, how much the radioactive metabolite contributes to measured radioactivity concentration in tissue). Helpful information about radioactive metabolites should be available from the early developmental stages of medicinal chemistry and tracer validation studies. However, this is often not considered as a major source of variability in SUV, even though, for some tracers, radiolabeled metabolites are likely a confound for SUV [28].

References

1. Gazdzinski S, Durazzo TC, Weiner MW, Meyerhoff DJ (2008) Are treated alcoholics representative of the entire population with alcohol use disorders? A magnetic resonance study of brain injury. Alcohol 42 (2): 67–76. doi:S07418329(08)000232 [pii]. https://doi.org/10.1016/j.alcohol.2008.01.002
2. Fein G, Landman B (2005) Treated and treatment-naive alcoholics come from different populations. Alcohol 36(2):19–26
3. Boileau I, Assaad JM, Pihl RO, Benkelfat C, Leyton M, Diksic M, Tremblay RE, Dagher A (2003) Alcohol promotes dopamine release in the human nucleus accumbens. Synapse 49(4): 226–231
4. Setiawan E, Pihl RO, Dagher A, Schlagintweit H, Casey KF, Benkelfat C, Leyton M (2014) Differential striatal dopamine responses following oral alcohol in individuals at varying risk for dependence. Alcohol Clin Exp Res 38(1):126–134. https://doi.org/10.1111/acer.12218
5. Urban NB, Kegeles LS, Slifstein M, Xu X, Martinez D, Sakr E, Castillo F, Moadel T, O'Malley SS, Krystal JH, Abi-Dargham A (2010) Sex differences in striatal dopamine release in young adults after oral alcohol challenge: a positron emission tomography imaging study with [(11)C]Raclopride. Biol Psychiatry doi:S0006-3223(10)00594-9 [pii]. https://doi.org/10.1016/j.biopsych.2010.06.005
6. Salonen I, Hietala J, Laihinen A, Lehikoinen P, Leino L, Nagren K, Ruotsalainen U, Oikonen V, Tuokkola T, Nanto V (1997) A PET study on the acute effect of ethanol on striatal D2 dopamine receptors with [11C] raclopride in healthy males. Hum Psychopharmacol 12:142–152
7. Volkow ND, Mullani N, Gould L, Adler SS, Guynn RW, Overall JE, Dewey S (1988) Effects of acute alcohol intoxication on cerebral blood flow measured with PET. Psychiatry Res 24(2):201–209
8. O'Connor S, Morzorati S, Christian J, Li TK (1998) Clamping breath alcohol concentration reduces experimental variance: application to the study of acute tolerance to alcohol and alcohol elimination rate. Alcohol Clin Exp Res 22(1):202–210
9. O'Connor S, Ramchandani VA, Li TK (2000) PBPK modeling as a basis for achieving a steady BrAC of 60 +/− 5 mg% within ten minutes. Alcohol Clin Exp Res 24(4):426–427
10. Ramchandani VA, O'Connor S, Blekher T, Kareken D, Morzorati S, Nurnberger J Jr, Li TK (1999) A preliminary study of acute responses to clamped alcohol concentration and family history of alcoholism. Alcohol Clin Exp Res 23(8):1320–1330
11. Plawecki MH, Boes J, Wetherill L, Kosobud AEK, Stangl BL, Ramchandani VA, Zimmermann US, Nurnberger JI Jr, Schuckit M, Edenberg HJ, Pandey G, Kamarajan C, Porjesz B, Foroud T, O'Connor S (2022) Binge and high-intensity drinking-associations with intravenous alcohol self-administration and underlying risk factors. Addict Biol 27(6):e13228. https://doi.org/10.1111/adb.13228
12. Stangl BL, Byrd ND, Soundararajan S, Plawecki MH, O'Connor S, Ramchandani VA (2022) The motivation for alcohol reward: predictors of progressive-ratio intravenous alcohol self-administration in humans. J Vis Exp 182. https://doi.org/10.3791/63576
13. Cyders MA, Plawecki MH, Corbin W, King A, McCarthy DM, Ramchandani VA, Weafer J, O'Connor SJ (2020) To infuse or ingest in human laboratory alcohol research. Alcohol Clin Exp Res 44(4):764–776. https://doi.org/10.1111/acer.14305
14. Cyders MA, Plawecki MH, Whitt ZT, Kosobud AEK, Kareken DA, Zimmermann US, O'Connor SJ (2021) Translating preclinical models of alcohol seeking and consumption into the human laboratory using intravenous alcohol self-administration paradigms. Addict Biol 26(6):e13016. https://doi.org/10.1111/adb.13016
15. Fang X, Deza-Araujo YI, Petzold J, Spreer M, Riedel P, Marxen M, O'Connor SJ, Zimmermann US, Smolka MN (2021) Effects of moderate alcohol levels on default mode network connectivity in heavy drinkers. Alcohol Clin Exp Res 45(5):1039–1050. https://doi.org/10.1111/acer.14602
16. Morzorati SL, Ramchandani VA, Li TK, O'Connor S (2002) Maintaining steady state arterial alcohol levels in rats by using a physiologically based pharmacokinetic model. Alcohol 28(3):189–195
17. Oberlin BG, Dzemidzic M, Eiler WJA 2nd, Carron CR, Soeurt CM, Plawecki MH, Grahame NJ, O'Connor SJ, Kareken DA (2018) Pairing neutral cues with alcohol intoxication: new findings in executive and attention networks. Psychopharmacology 235(9): 2725–2737. https://doi.org/10.1007/s00213-018-4968-7

18. Oberlin BG, Dzemidzic M, Tran SM, Soeurt CM, O'Connor SJ, Yoder KK, Kareken DA (2015) Beer self-administration provokes lateralized nucleus accumbens dopamine release in male heavy drinkers. Psychopharmacology 232(5):861–870. https://doi.org/10.1007/s00213-014-3720-1
19. Bragulat V, Dzemidzic M, Talavage T, Davidson D, O'Connor SJ, Kareken DA (2008) Alcohol sensitizes cerebral responses to the odors of alcoholic drinks: an fMRI study. Alcohol Clin Exp Res 32 (7): 1124–1134. doi:ACER693 [pii]. https://doi.org/10.1111/j.1530-0277.2008.00693.x
20. Yoder KK, Albrecht DS, Dzemidzic M, Normandin MD, Federici LM, Graves T, Herring CM, Hile KL, Walters JW, Liang T, Plawecki MH, O'Connor S, Kareken DA (2016) Differences in IV alcohol-induced dopamine release in the ventral striatum of social drinkers and nontreatment-seeking alcoholics. Drug Alcohol Depend 160:163–169. https://doi.org/10.1016/j.drugalcdep.2016.01.001
21. Yoder KK, Constantinescu CC, Kareken DA, Normandin MD, Cheng TE, O'Connor SJ, Morris ED (2007) Heterogeneous effects of alcohol on dopamine release in the striatum: a PET study. Alcohol Clin Exp Res 31(6): 965–973
22. Yoder KK, Kareken DA, Seyoum RA, O'Connor SJ, Wang C, Zheng QH, Mock B, Morris ED (2005) Dopamine D(2) receptor availability is associated with subjective responses to alcohol. Alcohol Clin Exp Res 29(6):965–970
23. Yoder KK, Kareken DA, Morris ED (2011) Assessing dopaminergic neurotransmission with PET: basic theory and applications in alcohol research. Curr Med Imaging Rev 7:118–124
24. Carson R (1993) PET parameter estimation using linear integration methods: bias and variability considerations. In: Uemura K, Lassen N, Jones T, Kanno I (eds) Quantification of brain function: tracer kinetics and image analysis in brain. Excerpta Medica, New York, pp 499–507
25. Carson RE (1996) Mathematical modeling and compartmental analysis. In: Harbert JC, Eckelman WC, Neumann RD (eds) Nuclear medicine: diagnosis and therapy. Thieme Medical Publishers, New York, pp 167–193
26. Jin X, Mulnix T, Planeta-Wilson B, Gallezot JD (1997) Carson RE (2009) accuracy of head motion compensation for the HRRT: comparison of methods. IEEE Nucl Sci Symp Conf Rec 2009:3199–3202. https://doi.org/10.1109/NSSMIC.2009.5401706
27. Innis RB, Cunningham VJ, Delforge J, Fujita M, Gjedde A, Gunn RN, Holden J, Houle S, Huang SC, Ichise M, Iida H, Ito H, Kimura Y, Koeppe RA, Knudsen GM, Knuuti J, Lammertsma AA, Laruelle M, Logan J, Maguire RP, Mintun MA, Morris ED, Parsey R, Price JC, Slifstein M, Sossi V, Suhara T, Votaw JR, Wong DF, Carson RE (2007) Consensus nomenclature for in vivo imaging of reversibly binding radioligands. J Cereb Blood Flow Metab 27 (9): 1533–1539. doi:9600493 [pii]. https://doi.org/10.1038/sj.jcbfm.9600493
28. Yoder KK, Territo PR, Hutchins GD, Hannestad J, Morris ED, Gallezot JD, Normandin MD, Cosgrove KP (2015) Comparison of standardized uptake values with volume of distribution for quantitation of [(11)C] PBR28 brain uptake. Nucl Med Biol 42(3): 305–308. https://doi.org/10.1016/j.nucmedbio.2014.11.003
29. Lopresti BJ, Klunk WE, Mathis CA, Hoge JA, Ziolko SK, Lu X, Meltzer CC, Schimmel K, Tsopelas ND, DeKosky ST, Price JC (2005) Simplified quantification of Pittsburgh compound B amyloid imaging PET studies: a comparative analysis. J Nucl Med 46(12): 1959–1972
30. Koole M, van Aalst J, Devrome M, Mertens N, Serdons K, Lacroix B, Mercier J, Sciberras D, Maguire P, Van Laere K (2019) Quantifying SV2A density and drug occupancy in the human brain using [(11)C]UCB-J PET imaging and subcortical white matter as reference tissue. Eur J Nucl Med Mol Imaging 46(2): 396–406. https://doi.org/10.1007/s00259-018-4119-8
31. Naganawa M, Gallezot JD, Finnema SJ, Matuskey D, Mecca A, Nabulsi NB, Labaree D, Ropchan J, Malison RT, D'Souza DC, Esterlis I, Detyniecki K, van Dyck CH, Huang Y, Carson RE (2021) Simplified quantification of (11)C-UCB-J PET evaluated in a large human cohort. J Nucl Med 62(3): 418–421. https://doi.org/10.2967/jnumed.120.243949
32. Endres CJ, Carson RE (1998) Assessment of dynamic neurotransmitter changes with bolus or infusion delivery of neuroreceptor ligands. J Cereb Blood Flow Metab 18(11):1196–1210
33. Morris ED, Fisher RE, Alpert NM, Rauch SL, Fischman AJ (1995) In vivo imaging of neuromodulation using positron emission tomography: optimal ligand characteristics and task length for detection of activation. Hum Brain Mapp 3:35–55

34. Morris ED, Yoder KK (2007) Positron emission tomography displacement sensitivity: predicting binding potential change for positron emission tomography tracers based on their kinetic characteristics. J Cereb Blood Flow Metab 27 (3):606–617. doi:9600359 [pii]. https://doi.org/10.1038/sj.jcbfm.9600359
35. Ramchandani VA, Umhau J, Pavon FJ, Ruiz-Velasco V, Margas W, Sun H, Damadzic R, Eskay R, Schoor M, Thorsell A, Schwandt ML, Sommer WH, George DT, Parsons LH, Herscovitch P, Hommer D, Heilig M (2010) A genetic determinant of the striatal dopamine response to alcohol in men. Mol Psychiatry doi: mp201056 [pii]. https://doi.org/10.1038/mp.2010.56
36. Constantinescu CC, Yoder KK, Kareken DA, Bouman CA, O'Connor SJ, Normandin MD, Morris ED (2008) Estimation from PET data of transient changes in dopamine concentration induced by alcohol: support for a non-parametric signal estimation method. Phys Med Biol 53(5):1353–1367
37. Yoder KK, Wang C, Morris ED (2004) Change in binding potential as a quantitative index of neurotransmitter release is highly sensitive to relative timing and kinetics of the tracer and the endogenous ligand. J Nucl Med 45(5):903–911
38. Cosgrove KP, Wang S, Kim SJ, McGovern E, Nabulsi N, Gao H, Labaree D, Tagare HD, Sullivan JM, Morris ED (2014) Sex differences in the brain's dopamine signature of cigarette smoking. J Neurosci 34(50):16851–16855. https://doi.org/10.1523/JNEUROSCI.3661-14.2014
39. de la Fuente-Fernandez R, Phillips AG, Zamburlini M, Sossi V, Calne DB, Ruth TJ, Stoessl AJ (2002) Dopamine release in human ventral striatum and expectation of reward. Behav Brain Res 136(2):359–363
40. de la Fuente-Fernandez R, Ruth TJ, Sossi V, Schulzer M, Calne DB, Stoessl AJ (2001) Expectation and dopamine release: mechanism of the placebo effect in Parkinson's disease. Science 293(5532):1164–1166
41. Kaasinen V, Aalto S, Nagren K, Rinne JO (2004) Dopaminergic effects of caffeine in the human striatum and thalamus. Neuroreport 15(2):281–285
42. Yoder KK, Kareken DA, Morris ED (2008) What were they thinking? Cognitive states may influence [(11)C]raclopride binding potential in the striatum. Neurosci Lett 430(1):38–42
43. Ceccarini J, Liu H, Van Laere K, Morris ED, Sander CY (2020) Methods for quantifying neurotransmitter dynamics in the living brain with PET imaging. Front Physiol 11:792. https://doi.org/10.3389/fphys.2020.00792
44. Alpert NM, Badgaiyan RD, Livni E, Fischman AJ (2003) A novel method for noninvasive detection of neuromodulatory changes in specific neurotransmitter systems. NeuroImage 19(3):1049–1060
45. Normandin MD, Morris ED (2006) Temporal resolution of ntPET using either arterial or reference region-derived plasma input functions. Conf Proc IEEE Eng Med Biol Soc 1: 2005–2008. https://doi.org/10.1109/IEMBS.2006.259716
46. Chumin EJ, Goni J, Halcomb ME, Durazzo TC, Dzemidzic M, Yoder KK (2018) Differences in white matter microstructure and connectivity in nontreatment-seeking individuals with alcohol use disorder. Alcohol Clin Exp Res 42(5):889–896. https://doi.org/10.1111/acer.13629
47. Durazzo TC, Abadjian L, Kincaid A, Bilovsky-Muniz T, Boreta L, Gauger GE (2013) The influence of chronic cigarette smoking on neurocognitive recovery after mild traumatic brain injury. J Neurotrauma 30(11):1013–1022. https://doi.org/10.1089/neu.2012.2676
48. Durazzo TC, Cardenas VA, Studholme C, Weiner MW, Meyerhoff DJ (2007) Non-treatment-seeking heavy drinkers: effects of chronic cigarette smoking on brain structure. Drug Alcohol Depend 87(1):76–82. https://doi.org/10.1016/j.drugalcdep.2006.08.003
49. Durazzo TC, Gazdzinski S, Banys P, Meyerhoff DJ (2004) Cigarette smoking exacerbates chronic alcohol-induced brain damage: a preliminary metabolite imaging study. Alcohol Clin Exp Res 28(12):1849–1860
50. Durazzo TC, Insel PS, Weiner MW, Alzheimer Disease Neuroimaging I (2012) Greater regional brain atrophy rate in healthy elderly subjects with a history of cigarette smoking. Alzheimers Dement 8(6):513–519. https://doi.org/10.1016/j.jalz.2011.10.006
51. Albrecht DS, Kareken DA, Yoder KK (2013) Effects of smoking on D/D striatal receptor availability in alcoholics and social drinkers. Brain Imaging Behav. https://doi.org/10.1007/s11682-013-9233-4
52. Poindexter EH Jr, Carpenter RD (1962) The isolation of harmane and norharmane from tobacco and cigarette smoke. Phytochemistry 1:215–221
53. Kettenes-van den Bosch JJ, Salemink CA (1977) Cannabis. XVI. Constituents of marihuana smoke condensate. J Chromatogr 131: 422–424

54. Novotny M, Merli F, Wiesler D, Saeed T (1982) Composition of the basic fraction of marijuana and tobacco condensates: a comparative study by capillary GC/MS. Chromatographia 15(9): 564–568
55. Sparacino CM, Hyldburg PA, Hughes TJ (1990) Chemical and biological analysis of marijuana smoke condensate. National Institute of Drug Abuse
56. ElSohly HN, ElSohly MA (2007) Marijuana smoke condensate: chemistry and pharmacology. In: ElSohly MA (ed) Forensic science and medicine: marijuana and the cannabinoids. Humana Press, Totowa, pp 67–96
57. Mawlawi O, Martinez D, Slifstein M, Broft A, Chatterjee R, Hwang DR, Huang Y, Simpson N, Ngo K, Van Heertum R, Laruelle M (2001) Imaging human mesolimbic dopamine transmission with positron emission tomography: I. accuracy and precision of D (2) receptor parameter measurements in ventral striatum. J Cereb Blood Flow Metab 21(9): 1034–1057
58. Ichise M, Liow JS, Lu JQ, Takano A, Model K, Toyama H, Suhara T, Suzuki K, Innis RB, Carson RE (2003) Linearized reference tissue parametric imaging methods: application to [11C]DASB positron emission tomography studies of the serotonin transporter in human brain. J Cereb Blood Flow Metab 23(9): 1096–1112. https://doi.org/10.1097/01.WCB.0000085441.37552.CA

Chapter 7

Oral Alcohol Administration Methods for Alcohol Research

Melissa A. Cyders

Abstract

Oral alcohol administration methods are a common way to assess alcohol intake in humans within a controlled, laboratory setting. A variety of considerations and methods exist to modify oral alcohol administration for different research goals and questions. In this chapter, I describe methodological considerations for conducting oral alcohol administration studies, including strengths and weaknesses of this approach and questions appropriate for examination with these techniques.

Key words Oral alcohol, Alcohol challenge, Alcohol administration, Alcohol use disorder, Experimental, Laboratory

1 Introduction

There are two main ways of assessing alcohol intake in the human laboratory: Oral alcohol administration or intravenous alcohol administration [1] (see a review by Cyders, Plawecki, et al.). The latter is discussed in a separate chapter in this collection. The goal of this chapter is to discuss how and why researchers use *oral* alcohol administration in human laboratory studies of alcohol risk and outcomes.

A common method of assessing how much alcohol one is consuming is to ask them. These measures of alcohol intake can take many forms, including briefly asking for one's general use (e.g., asking them how many days, in a typical week or month, one consumes alcohol and how many drinks they tend to consume in a week), screening briefly for risky alcohol use (e.g., asking how many times in the past year the person has engaged in binge drinking), asking for one's experience with alcohol and its related problems, for example, using the Alcohol Use Disorders Identification Test [2], or asking for one's use (e.g., number of drinks) for each day over a preset number of days (e.g., The Timeline Followback [3]). Although these measures have clear merit, in that they

Melissa A. Cyders (ed.), *Translational Research Methods for Alcohol Use Disorders*, Neuromethods, vol. 201,
https://doi.org/10.1007/978-1-0716-3267-3_7,

characterize one's use in general, in real life, over a longer period than is often studied in a laboratory-based study, there have been questions as to how accurate one can be when self-reporting their alcohol use, given that individuals may try to hide the true extent of their drinking, or may not accurately remember given that alcohol affects memory formation.

Thus, as a complementary and better-controlled procedure, researchers began to assess alcohol consumption in a human laboratory setting, in order to ensure more precise and accurate measurement of the amount of alcohol one consumes, and the specific, objective effects of such alcohol consumption, without the limitations of one's self-reporting. Not surprisingly, since the most common route of administration for alcohol is the oral route, most studies have used oral alcohol paradigms, with a huge impact on our understanding of risk factors and outcomes associated with alcohol use. Although a full review of this literature is outside the scope of this chapter, I mention one prime example: The use of oral alcohol administration in the laboratory has helped to better understand how alcohol is experienced by individuals, and how this experience may be biased by and predict one's alcohol use disorder risk. A review [4] summarized this literature and characterized the effects of orally administered alcohol on subjective responses to alcohol (i.e., feelings of intoxication, high, sedation in response to alcohol), finding support for the idea that individuals with a family history of alcoholism may show an initial low level of response to alcohol's effects following consumption (termed the Low Level of Response Model; [5]) that may predict later risk for alcohol use disorder. Additionally, one study [6] utilized a laboratory paradigm to measure one's initial response to alcohol consumption and was able to predict alcohol use disorder symptoms 6 years later. These studies together support the use of alcohol challenge paradigms in better understanding alcohol use phenotypes and experiences.

Oral alcohol administration approaches take two main forms: the first is an *alcohol challenge study*, where a researcher pre-determines a targeted dose of alcohol to be administered to a participant, under controlled setting, and then measuring the participants response to the alcohol administration in some way (e.g., reaction time, risk-taking, mood, etc). For this approach, the alcohol is best considered as an independent variable (or a moderator) that is having an effect on a later dependent variable. The second is an *alcohol self-administration approach*, where participants are free to drink as much or as little of a pre-determined amount of alcohol provided by the researcher over a set period. For this approach, the alcohol is best considered a dependent variable, with another factor influencing how much alcohol one consumes, or how quickly they consume it. Since the materials, methods, and questions vary across these two paradigms, this chapter treats these separately.

2 Oral Alcohol Challenge (Experimenter-Administered Approach)

1. *Materials:* Oral alcohol challenge studies require alcohol, usually some form of distilled spirits or grain alcohol, that is mixed with a non-alcoholic drink (e.g., juice, sweetened or unsweetened drink). In addition to these materials, such experiments require a safe, comfortable laboratory space, with access to a private bathroom, technicians with the expertise to calculate and administer alcohol doses safely and to monitor participants for negative reactions to such alcohol administration, and minor materials such as glassware, scales and cylinders, breath alcohol meters, etc.
2. *Methods:* Doses of alcohol are scaled to the participant's weight and sex (see a useful program for such calculations by [7]) to target a particular average peak breath alcohol concentration (BrAC). The beverage is then required to be drunk by the participant within a set period of time (e.g., 30 min) in an attempt to control the resulting timing of the BrAC across participants, allowing for more standardized measurement of outcome variables, whatever they may be, at a specific time and point in the BrAC curve. Often, researchers will measure the participant's BrAC at specific times following ingestion and will time the measurement of outcome variables during the ascending limb, peak, and/or descending limb of BrAC.
3. *Questions:* Since alcohol consumption is somewhat standardized in alcohol challenge approaches, these approaches are best used when studying alcohol as an independent variable. For example, a researcher can use this approach to study the effect of different levels of alcohol ingestion, BrAC, and timing of this ingestion on outcomes, including driving (e.g., [8]), inhibitory control (e.g., [9]), or economic demand and craving (e.g., [10]). Some have used such alcohol challenge approaches as a way of measuring how one's initial reactions to alcohol may predict subsequent alcohol risk or consumption (e.g., [11]) or to assess how one's response to an alcohol challenge may be influenced pre-existing risk factors (e.g., family history of alcoholism; [12]).

   Additionally, questions that can be answered using a relatively low dose of alcohol are best answered using an alcohol challenge approach. Target BrACs for oral alcohol challenge studies usually range from 0 mg/dl (i.e., a placebo dose) to ~80 mg/dl, although there are some researchers using higher target BrACs (e.g., [10]) and one study was able to target 120 mg/dl without adverse effects [13].

3 Oral Alcohol Self-Administration (Participant-Administered Approach)

1. *Materials:* Oral alcohol self-administration studies require some form of alcohol, with a broader range typically used in these studies as compared to alcohol challenge studies. In addition, such experiments require a safe, comfortable laboratory space, with access to a private bathroom, technicians with the expertise to calculate and administer alcohol safely and to monitor participants for negative reactions to such alcohol administration, and minor materials such as glassware, scales and cylinders, breath alcohol meters, etc.
2. *Methods:* In oral alcohol self-administration studies, participants are allowed to drink as much or as little of an alcoholic drink as they would like within a predetermined amount of alcohol provided and within a predetermined amount of time. Additionally, at times, participants may choose between consuming the alcohol or choosing a different reinforcer, such as money or food. The choice of which alcoholic beverage to use in oral alcohol self-administration studies is usually made to maximize the likelihood that the individual will drink alcohol similarly to how they would outside of the lab, and may include choosing the participant's most frequently consumed alcoholic beverage, or choose a type of beverage (e.g., beer) that is then matched to the recruitment sample (i.e., sampling people who like to drink beer). Participants may undergo methods and procedures prior to the opening of the self-administration period, such as a priming dose (i.e., the participant is asked to consume one drink that is experimenter-administered), a mood induction, or other manipulation of interest. Then, a period of ad libitum alcohol consumption is initiated (usually between 1 and 2 h), where the individual is free to choose how much or how little alcohol to consume. Other tasks (e.g., a taste rating form for the different alcohol beverages that are offered) may be administered during the ad libitum period. The outcome of interest in these types of studies is often the amount of alcohol (e.g., in mL) consumed or the timing of the consumption throughout the ad libitum period.
3. *Questions:* Because alcohol consumption (either rate or amount) is often used as the dependent variable in oral alcohol self-administration studies, this type of design is best suited to studies seeking to document the determinants or moderators of alcohol consumption. For example, researchers may administer a mood induction (e.g., [14]) or stress task [15] to see whether such a task influences the rate or amount someone consumes alcohol or they may administer a medication or other manipulation designed to reduce alcohol consumption.

Whether a manipulation can be repeated without learning effects determines whether such approaches are done in a between or within subjects design, but both are conceptually possible.

Strengths There are a number of strengths in using oral alcohol paradigms. First, oral alcohol challenge approaches better ensure that all participants are ingesting a similar dose of alcohol, and are approaching a better controlled BrAC, than would occur in the natural environment, thus allowing to examine alcohol's effects at a particular level or timing in the BrAC curve across participants. Additionally, administering alcohol to participants in a controlled way and measuring a behavioral outcome removes any self-report bias concerning participants' reporting of how much alcohol they consumed or how they experienced the effects of such alcohol in a "real-world" setting, as such reports are likely limited by one's openness, memory, awareness, and willingness to report. Third, this approach is relatively low risk, as it is only conducted in non--naïve alcohol-drinkers who report consuming doses similar to those given in the laboratory setting, thus reducing adverse effects of the administration. Fourth, this approach uses an ecologically valid route of administration and relatively little technician expertise and cost. Fifth, this approach allows great flexibility to control or vary beverage characteristics to meet the specific needs of the experiment, including, but not limited to, using a placebo, using a novel beverage, or using a beverage for which the individual has a long consumption history. This approach also easily allows for modifying other experimental environment characteristics, such as the presence of other peers or drinkers. Finally, oral alcohol challenges allow characterizing relationships between alcohol consumption and neurophysiological outcomes (e.g., electrophysiological responses or striatal dopamine release) that would not be possible using self-report alcohol approaches. Thus, oral alcohol ingestion approaches offer a number of strengths that make them well-used and important for the field.

Limitations A number of limitations exist for oral alcohol approaches. The biggest limitation is that, despite attempts to scale one's alcohol dose based on personal characteristics, resulting BrAC peaks and curves still vary significantly across individuals, contributing to increased error variance in experimental outcomes (see Cyders et al., 2020) that can be better controlled in approaches such as intravenous alcohol. Some of this variability may be addressed via statistical controls or covariates, but it is likely that it can never be completely eliminated using oral approaches. A second limitation is that they have been criticized as not being ecologically valid, as the alcohol consumption takes place in a laboratory space that is often quite different from the location that the individual most commonly consumes alcohol, often the individual is alone or at least not with the individuals with whom they most often consume alcohol, or they may be asked to consume a type or form of alcohol that they have not consumed before. Some of these criticisms can be addressed with solid study design, including making laboratory spaces mimic a bar or college dorm room, allowing other drinkers or friends to attend sessions, or having the person consume their most frequently consumed alcoholic beverage. However, it should be noted that there is always a trade-off between experimental control and ecological validity; thus, attempts to maximize ecological validity may limit experimental control or introduce error variance into the experiment. Third, it is difficult to ensure that there are not factors outside of the laboratory that influence how much a person consumes in an oral alcohol self-administration paradigm; for example, some participants have reported not drinking so that they can leave the session early or because they have a test or have to work the day following the session. Although experimenters attempt to control for these with study policies and participant instructions, they cannot be completely removed. Finally, laboratory assessments of drinking represent one's drinking at one point in time and may not always generalize to one's drinking in their "real life" over longer periods of time.

4 Conclusions

In conclusion, oral alcohol administration techniques are flexible techniques to answer a number of important questions about the causes and effects of alcohol use in humans. They are well suited to laboratory study and are more controlled than self-reports of alcohol use and consumption. These approaches impart control on one's level, timing, and dose of alcohol consumed, creating better comparison across subjects, and use an ecologically valid route of administration. However, they cannot control for differences across and within individuals on alcohol metabolism or external unknown effects that may impact drinking in the laboratory, nor can laboratory-based studies of alcohol consumption be called ecologically valid in the truest sense. Overall, these methods are important in the study of human alcohol use and may be particularly valuable with compared and combined with alternative approaches that can overcome the limitations of these approaches, such as intravenous alcohol administration (see Plawecki et al. in this collection) and ecological momentary assessment approaches.

References

1. Cyders MA, Plawecki MH, Corbin W, King A, McCarthy DM, Ramchandani VA, Weafer J, O'Connor SJ (2020) To infuse or ingest in human laboratory alcohol research. Alcohol Clin Exp Res 44(4):764–776
2. Saunders JB, Aasland OG, Babor TF, De La Fuente JR, Grant M (1993) Development of the alcohol use disorders identification test (AUDIT): WHO collaborative project on early detection of persons with harmful alcohol consumption. Part 11. Addiction 88:791–804
3. Sobell LC, Sobell MB (1992) Timeline followback. In: Measuring alcohol consumption. Humana Press, Totowa, NJ, pp 41–72
4. Quinn PD, Fromme K (2011) Subjective response to alcohol challenge: a quantitative review. Alcohol Clin Exp Res 35(10): 1759–1770
5. Schuckit MA (2009) An overview of genetic influences in alcoholism. J Subst Abus Treat 36:S5–S14
6. King AC, McNamara PJ, Hasin DS, Cao D (2014) Alcohol challenge responses predict future alcohol use disorder symptoms: a 6-year prospective study. Biol Psychiatry 75(10):798–806
7. Curtin JJ, Fairchild BA (2003) Alcohol and cognitive control: implications for regulation of behavior during response conflict. J Abnorm Psychol 112:424–436
8. Fillmore MT, Blackburn JS, Harrison EL (2008) Acute disinhibiting effects of alcohol as a factor in risky driving behavior. Drug Alcohol Depend 95(1–2):97–106
9. Weafer J, Fillmore MT (2016) Low-dose alcohol effects on measures of inhibitory control, delay discounting, and risk-taking. Curr Addict Rep 3(1):75–84
10. Amlung M, McCarty KN, Morris DH, Tsai CL, McCarthy DM (2015) Increased behavioral economic demand and craving for alcohol following a laboratory alcohol challenge. Addiction 110(9):1421–1428
11. Li J, Murray CH, Weafer J, de Wit H (2020) Subjective effects of alcohol predict alcohol choice in social drinkers. Alcohol Clin Exp Res 44(12):2579–2587
12. Newlin DB, Thomson JB (1990) Alcohol challenge with sons of alcoholics: a critical review and analysis. Psychol Bull 108(3):383
13. Vena A, King AC (2019) Testing the allostasis theory in light, heavy, and AUD drinkers. Alcohol Clin Exp Res 43:239A
14. Cyders MA, Zapolski TC, Combs JL, Settles RF, Fillmore MT, Smith GT (2010) Experimental effect of positive urgency on negative outcomes from risk taking and on increased alcohol consumption. Psychol Addict Behav 24(3):367
15. Thomas S, Bacon AK, Sinha R, Uhart M, Adinoff B (2012) Clinical laboratory stressors used to study alcohol–stress relationships. Alcohol Res 34(4):459

Chapter 8

Intravenous Alcohol Administration Studies in Humans: Investigator-Prescribed and Self-Administration Paradigms

McKenzie R. Cox, James M. Hays, Tiffany M. Bravo, Ann E. K. Kosobud, Bethany L. Stangl, Vijay A. Ramchandani, Sean O'Connor, and Martin H. Plawecki

Abstract

Alcohol administration experiments, or sometimes termed alcohol challenges, fall into two classes: those in which the investigator determines the dose or exposure of alcohol to be delivered and those in which the participant determines it. Investigator-administration paradigms are used in both preclinical and human studies and seek to inform the impact of differential sensitivity to prescribed level and duration of alcohol concentrations, and to its rate of change with respect to time (slope), on physiological, behavioral, and subjective responses to alcohol. Self-administration paradigms represent the gold standard for assessing drug seeking and consumption behavior in animal research with increasing application to human laboratory assessment. This review covers both classes of studies in humans, but focuses on the methods using an intravenous route of administration using the Computer-assisted Alcohol Infusion System (CAIS). The technique couples a physiologically based pharmacokinetic model of alcohol distribution and elimination with parameters tailored to each participant to overcome the two- to threefold range of peak breath alcohol concentration (BrAC) and time to peak BrAC after oral alcohol administration. The procedures to accomplish both experimental types are provided.

Key words Alcohol clamp, Alcohol self-administration, Operant techniques, Progressive ratio, Free-access, Fixed ratio

1 Introduction

Alcohol administration experiments, or sometimes termed alcohol challenges, fall into two classes: those in which the investigator determines the dose or exposure of alcohol to be delivered and those in which the participant determines it. This review covers both classes but focuses on the methods using an intravenous route of administration.

The authors "McKenzie R. Cox" and "James M. Hays" are co-first authors.

Melissa A. Cyders (ed.), *Translational Research Methods for Alcohol Use Disorders*, Neuromethods, vol. 201,
https://doi.org/10.1007/978-1-0716-3267-3_8,

Investigator-prescribed administration (IA) paradigms are used in both preclinical and human studies and seek to inform the impact of differential sensitivity to prescribed level and duration of alcohol concentrations, and to its rate of change with respect to time (slope), on physiological, behavioral, and subjective responses to alcohol. Acute (within-session) tolerance of these responses to alcohol is another phenomenon of interest, as is the effect of first-pass metabolism.

Self-administration (SA) paradigms represent the gold standard for assessing drug seeking and consumption behavior in animal research. Alcohol SA paradigms discriminate groups selectively bred for alcohol preference and for genetic and phenotypic determinants of alcohol use disorder (AUD) risk and have been used to screen substances for both abuse liability and treatment efficacy [1–6]. Thus, evaluation of potential clinical treatments for AUD is informed by translation of these animal self-administration paradigms.

Oral alcohol ingestion is the most used approach for human studies, typically delivered as a predetermined dose of alcohol, calculated to achieve a target alcohol exposure, mixed at various concentrations in a variety of vehicles, and often conducted in a "laboratory bar setting." Unfortunately, using ingestion to investigate the effects of alcohol administration consistently across men and women is challenging: a two- to threefold range of peak breath alcohol concentration (BrAC) and time to peak BrAC ensues (*see* Fig. 1, top) both within and between participants (*see* [7–9] amongst many others). BrAC corresponds to arterial alcohol concentration, the concentration to which the brain is exposed [10]. Due to uncontrollable absorption kinetics, this substantial variation in brain alcohol exposure with oral dosing cannot be overcome, even when dosing is adjusted for body weight, lean body mass, and the volume of distribution (*see* [11–16] among many other studies). Thus, the response following oral alcohol administration becomes a unique, potentially rewarding, or aversive experience, complicating at best and confounding at worst interpretation of the pharmacodynamic measures obtained from these paradigms. Safety considerations also complicate oral alcohol self-administration techniques. Investigators must limit dosing to prevent dangerously high peak values of BrAC in the participants who most rapidly absorb the maximum allowable dose. Consequently, these experiments limit the amount of available alcohol, introducing a ceiling effect (for example, [17]) or resulting in peak concentrations likely well below that for normal drinking [18–20]. Intravenous (IV) alcohol administration paradigms offer an alternative to the oral challenge methodology because absorption kinetics are no longer a constraint.

Several alcohol infusion techniques have been utilized to overcome these challenges [21–30]. The most commonly used IV

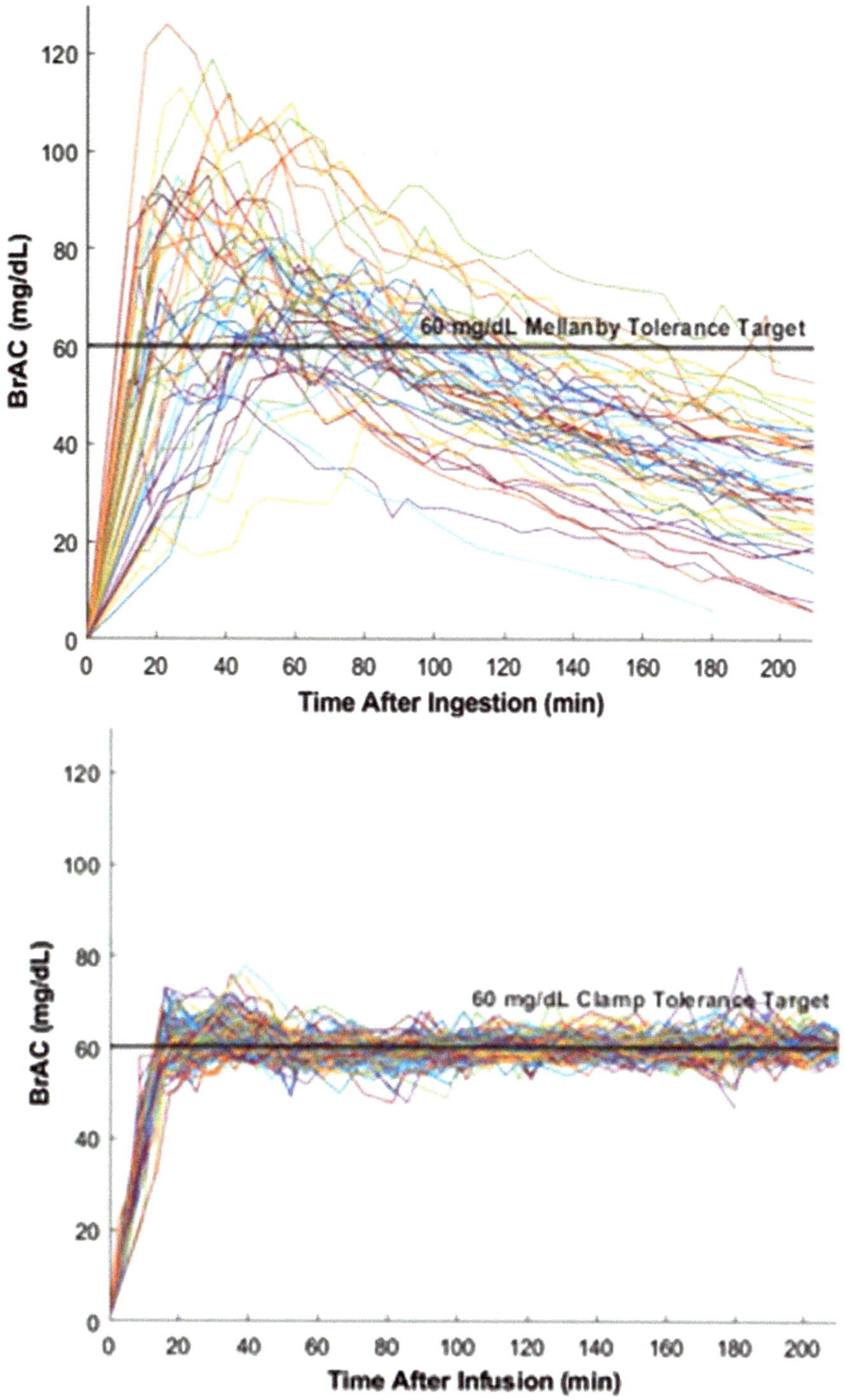

Fig. 1 Breath Alcohol Concentration (BrAC) variability for the measurement of Acute Tolerance. Top: Variability in BrAC exposure trajectories from a controlled oral dose of alcohol for measurement of tolerance at 60 mg/dL (80 mg/dL peak target, $n = 44$). (Data from [7].) Bottom: Variability in breath alcohol concentration during a 3-h 60 mg/dL BrAC clamp experiment ($n = 130$). (Dataset from [8, 9])

alcohol technique couples a physiologically based pharmacokinetic (PBPK) model of alcohol distribution and elimination [31–33] with parameters tailored to each participant [34]. This combination is embedded in a software program called the Computer-assisted Alcohol Infusion System (CAIS) [35–37] that directly controls the infusion pump rate delivering 6% ethanol in a saline vehicle to the participant. The key innovation with CAIS methodology is that dosing, as the input variable, is replaced with the desired time course of the brain's exposure to alcohol. The technique minimizes the variability in exposures during delivery of the alcohol exposure (*see* Fig. 1, bottom). The route of delivery is not a constraint since the subjective response to similar exposures resulting from intravenous and ingested alcohol are similar and non-differentiable [38].

CAIS enables safe brain exposures to alcohol commensurate with those achieved by actual drinking habits. Since PBPK modeling predicts the entire future trajectory of BrAC resulting from a given reward, infusions that would exceed a pre-determined BrAC safety limit can be prevented. Infusion also enhances safety should the participant experience an untoward event or the experiment needs to be stopped. With ingestion, alcohol in the gastrointestinal tract continues to be absorbed, but IV administration eliminates this reservoir of alcohol, so that discontinuation of the infusion always results in an immediate decrease in BrAC. Thus, a wide dynamic range of safe, achievable BrAC levels is available with intravenous alcohol self-administration paradigms. For a more detailed contrast of intravenous and oral alcohol challenge techniques, please see [39].

The original IA class CAIS paradigm was the alcohol clamp (*see* Fig. 1, bottom). The clamp's desired time course of BrAC begins with a linear ascent to a target concentration in a prespecified interval (T1), followed by maintenance of the target concentration for as long as the investigator specifies. The target concentration is typically 60 mg/dl (high enough to reveal family history differences in dependent measures), but any concentration is possible, and many experiments have been performed at a target concentration in the range of 80–120 mg/dL. Measurements of dependent variables (for example, subjective perceptions, cognitive skills, eye movements, psychophysiological, neuroendocrine) are obtained at baseline, immediately after T1 and periodically during the clamped interval. The difference between baseline and T1 measurements establishes the participant's initial sensitivity to alcohol, and the time course of measurements during the clamp quantifies their acute adaptation to alcohol. The alcohol clamp has been used to study how subjective and physiological measures relate to family history of AUD [40–42], genetic variants [8, 43–46], recent drinking history [8, 47], endocannabinoid levels [48], cognitive functioning, and the role of acetaldehyde [49], neural connectivity patterns [50], as well as other risks, including low inhibitory control [9, 51, 52] and alteration of atrial electrophysiology [53]. The infusion rate profile delivered during the clamp has also been used to study metabolism [43, 45, 54–56], calculating the participant's alcohol elimination rate in gm/hr. The alcohol clamp has also been used in translational animal studies since, with appropriately scaled parameters, the PBPK model works across species (for example, [57, 58]). Other IA prescribed BrAC trajectories can be used to investigate the participant's sensitivity to the slope of BrAC [59, 60] or to a trajectory mimicking the individual participant's own BrAC trajectory to an oral challenge or to the average trajectory of many participants [7, 38].

For the SA class of paradigms, the desired BrAC trajectory is an incremental addition (or subtraction) from the currently established trajectory. Preclinical literature suggests that alcohol seeking and intake are separable and complementary constructs [61–66], informing different aspects of drinking. Progressive-ratio paradigms are thought to reflect motivation for acquiring a specific reward and require increasing "work" to earn the next return. Free-access paradigms are thought to reflect the rewarding aspects of the drug itself, and alcohol can be consumed ad lib in a specified interval, subject to safety constraints. Importantly, studies in laboratory animals have shown that both alcohol seeking and motivation are affected by abstinence [67–69] and are differentially responsive to modulation, including by intervention with putative therapeutics [63, 70–72]. Consequently, efforts have been made to translate these techniques to the human laboratory setting.

Using oral alcohol administration techniques, most human alcohol self-administration experiments have employed free-access (FA) paradigms; participants may orally consume alcohol containing beverages in a specified interval which may be preceded by a prime. Investigators have validated a preference for alcohol over placebo [73], examined the relationship between the number of alcohol rewards and subjective effects [74], compared the rewarding aspects of alcohol vs. diazepam [75], identified effects of naltrexone [19, 76, 77] as well as stress [78], and examined differences attributable to family history [79, 80], among others.

The first CAIS-based intravenous alcohol SA techniques implemented a free-access paradigm [35] and then reported family history effects [36]. Subsequent investigations reported interactions between alcohol self-administration and both mood and gender [81] as well as negative urgency [82] on peak BrAC, influences of expectancies and impulsivity on the exposure-response relationship [83], suggestibility [84], the role of impaired control [85], sex effects in adolescents [86], the relationship between craving and stable FA alcohol exposures [87], the determinants of hangover [88], relationships between rate of self-exposure and alcohol use disorder risk [89, 90], and lack of association between OPRM1 A11G polymorphism and ASA outcomes [91]. Most recently, a "rate control" IV alcohol free-access type technique was described. The approach takes further advantage of the precise control allowed by pharmacokinetic modeling and within-session feedback. In the rate control paradigm, participants are still free to administer intravenous alcohol ad libitum, but they choose the rate of change (i.e., slope) of the next 3-min segment of the experiment [92]. In contrast to other infusion paradigms, where participants experience identical incremental alcohol rewards, here each incremental exposure is under the participant's control and aims to discriminate high-intensity risk drinking patterns.

A few human studies have employed Progressive Ratio (PR) self-administration paradigms to quantify motivation to seek alcohol under the effects of various modulators [93–97]; all are designed to identify the point at which earning the next reward is no longer worth the effort. As successive rewards require an increasing amount of work, the work schedule can be designed to limit a participant's ability to raise or even maintain their BrAC over the course of the experimental session. Human alcohol studies employing a PR schedule have mostly used oral alcohol self-administration methods; examining the effect of naltrexone [97], as well as the role of dopamine [96] and nicotine [95] in alcohol self-administration. Within CAIS, currently available work is one of two types: simple button presses [98, 99] or the Constant Attention Task (CAT) [37]. The CAT is an operant task requiring participants to attend and respond to a stimulus in a specified time period. As detailed below, the difficulty of the task adjusts so that fatigue or intoxication do not affect the success rate across the experiment. Investigators utilizing intravenous alcohol PR paradigms have reported sensitivity to pharmacogenomic interactions in a pilot study [37], an interaction of gender and mood on peak alcohol exposure [81], sex difference in the response to abstinence [100], family history effects on motivation for alcohol [87], both test-retest reliability and strong association between recent drinking history and self-administration measures [99], and the role of ghrelin on motivation for alcohol [98]. Finally, other investigators have combined methods, using the alcohol clamp as a priming exposure upon which alcohol-self-administration experiments have been completed (for example, [101]).

2 Materials

CAIS is optimally designed to be used with computer-controllable infusion pumps so that, with measured BrAC entries, the embedded control algorithm will re-compute the infusion rate required to match the desired incremental or overall exposure trajectory. As currently configured, CAIS updates the infusion rates to an attached infusion pump every 30 s. A computer-controlled infusion pump is not strictly required, but in its absence, the fidelity to the desired trajectory depends upon regular manual adjustments of the infusion rate based upon CAIS predictions.

For both safety and to achieve and maintain fidelity to the desired exposure profile, a calibrated breath alcohol meter with disposable mouthpieces is needed to periodically check the participant's BrAC. Compensating for non-linearities in the meter's performance are important to safety when BrACs above 150 mg/dL are planned, and CAIS incorporates them automatically.

The use of a digital refractometer to estimate the alcohol infusate concentration is also recommended, but not necessarily required if care is taken in its preparation.

2.1 Durable Equipment and Software

- Computer with updated Windows 10 operating system, administrative privileges, Microsoft Office, 3 USB ports.
- Computer-controlled infusion pumps:

  iMed Gemini PC-2TX infusion pump(s).

  The procedures below will focus on procedures for the iMED Gemini PC-2TX infusion pumps.

  Braun Vista Infusomat fmS Perfusor type 8715424 pump (s) (available in Germany and other European countries).

  Note: A driver for the Harvard Apparatus P-1500 Peristaltic pump is under development. Drivers for the since-recalled BodyGuard Twins infusion pumps also exist.
- CAT6 Ethernet Cable.
- Programmable Infusion Pump Computer Control-Related Material for the iMed Gemini PC-2TX:
  - USB to RS232 DB9 Serial Adapter Cable—M/M (1 or 2 ports, depending on need for alternative reward).
  - DB9 to RJ45 Male-to-Female Modular Serial Adapter.
  - DB9 to RJ45 Female-to-Female Modular Serial Adapter.
- Griffin Powermate buttons or similar USB-powered buttons (for progressive work SA paradigms).
- USB port multiplier (for SA paradigms).
- Digital refractometer (Recommended).
- Breath Alcohol Meter (breathalyzer).
- Participant video monitor with cabling and port to match com puter configuration.

  Required for alcohol SA experiments and other designs where direct engagement with the participant is desired.

If two external monitors are used, it is strongly recommended that matching resolutions are set to minimize distortion across displays.

- CAIS computer software package.
- Microsoft Office, including Access. CAIS utilizes Microsoft Access to operate experiments and adjust experiment settings.
- Microsoft .netframework.
- Optional: Matlab.

2.2 Consumable Materials

- Intravenous catheters: 20-gauge or larger.

  For example: 20GA 1.00IN 1.1 x x25mm, ref 383516, catheter
- Alcohol cleaning pads (70% Isopropyl Alcohol).
- Infusion sets compatible with the Infusion pump.

  For example: CareFusion SmartSite Infusion Set 2420-0500 for the iMed Gemini PC-2TX
- Leuer Lock extension sets for interconnection between multiple bags of alcohol infusate.

  For example: Baxa extension set, male-to-male Leuer Lock, ref 87
- Infusion extension sets may be required depending on the physical dimensions of the experimental environment.
- Breathalyzer mouthpieces.

2.3 Alcohol Infusate

The quantity of alcohol infusate required varies according to the experimental design (target concentration, duration, etc.) and participant characteristics (larger total body water, i.e., a larger volume of distribution of alcohol requires a larger volume of alcohol infusate), but generally falls within a range of 0.5–2 L. Depending on local regulations, the alcohol infusate may need to be prepared from sterile constituents by the investigator or within a pharmacy setting. At our institutions, an approximately 6% v/V alcohol infusate is prepared using sterile compounding techniques using 95% v/v alcohol for injection in 1/2 normal saline (0.45% sodium chloride injection USP). The choice of vehicle varies somewhat across location and time, with experimental use of both normal saline and Lactated Ringers solutions. Alcohol is highly osmolar and any infusate containing it will also be highly osmolar (but dramatically less than what is consumed orally).

For any alcohol self-administration studies that include saline as an optional alternative reward (for example, free-access or progressive ratio SA designs), a supply of sterile normal saline (0.9% sodium chloride injection USP) is also required.

3 Methods

3.1 Initial Equipment and Software Setup

3.1.1 Initial CAIS Computer Setup

1. CAIS software has a relatively low impact on computer operating resources, but it is ideal to minimize background processes to reduce the likelihood of experimental interference.
   (a) Turn off notifications.
   (b) Minimize background applications: In Task Manager, Start-Up tab, disable Automatic Start-Up of any applications or programs unnecessary for basic functionality or experimental requirements, for example, Adobe software, internet browsers, Skype, MS Teams, etc.
   (c) Modify Power Options. In Control Panel, Power Options, Plan Settings, set computer to Never Turn Off Display or Put Computer to Sleep. Then, in Advanced Power Settings, adjust the Time to Turn Off the Hard Disk to a very high number (for example, 99,999), and ensure that the rest of the power settings are set to never turn off components.
   (d) Update computer and software.
   (e) Review Scheduled Tasks. Many scheduled tasks, such as antivirus software scans, are resource-intensive and could interfere with an experiment. Ensure that any scheduled processes occur when the computer is not in use running an experiment.
2. Install USB to RS232 DB9 Serial Adapter Cable for Gemini PC-2TX
   (a) Insert the USB end of the cable into an available USB port on the computer. While other components will correctly function with a port adapter, it is ideal for the USB to RS232 DB9 Serial Adapter Cable to connect **directly to a computer USB port**.
   (b) Com ports are defined within CAIS experimental settings and must match system settings for proper infusion pump control. Open Device Manager and expand Ports (COM & LPT). Right click and open properties on the USB to RS232 DB9 Serial Adapter listing. Click Port Settings, Advanced, and change the port to COM11, and then click OK and close the properties.
      1. If running a two-fluid experiment with two RS232 ports, right click and open properties for the other Serial Adapter listing and repeat the process, selecting COM12 as the port number.
   (c) The adapter cable is installed and ready to run an experiment (it can be disconnected between experiments).

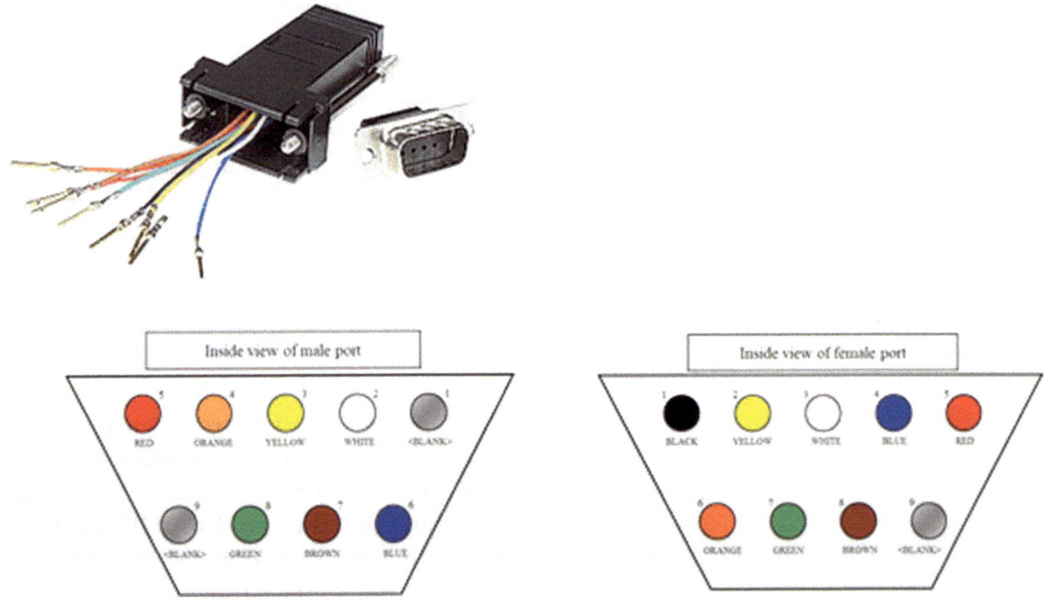

Fig. 2 DB9 to RJ45 modular serial adapters and pin diagrams. StarTech.com® adapters are displayed. https:/www.startech.com/en-us/cables/gc98mf;https:/www.startech.com/en-us/cables/gc98ff

(d) Place a label on the USB port and when setting up experiments, always insert the cable into the same USB port on the computer. *If it is necessary to use a different USB port, the port settings described above need to be mapped again.*

(e) Prepare the DB9 to RJ45 Modular Serial Adapters by wiring the adapters, as shown in Fig. 2.

(f) Cable Connections. Connect (a) the DB9 to RJ45 Female-to-Female Modular Serial Adapter to (b) the RS232 port of the USB to RS232 DB9 Serial Adapter, and then connect (c) the CAT6 Ethernet Cable to the DB9 to RJ45 Female-to-Female Modular Serial Adapter (a) on one end, and (d) the DB9 to RJ45 Male-to-Female Modular Serial Adapter. Connect (d) the DB9 to RJ45 Male-to-Female Modular Serial Adapter to the RS232 port in the back of (e) the Gemini PC-2TX pump. Connect the USB end of the adapter to (f) the computer. The connections are depicted in Fig. 3.

3. Install Peripheral Components

(a) Connect the USB port multiplier (if needed) and ensure the driver installs correctly. Attach other components to the port multiplier, assuring that each of the drivers install and the components function correctly (such as a camera, number pad, and response buttons). In most cases, Windows will select and install the correct drivers for peripheral components automatically. In rare cases, the user may need to manually install a driver, in which case they should follow manufacturer's recommendations.

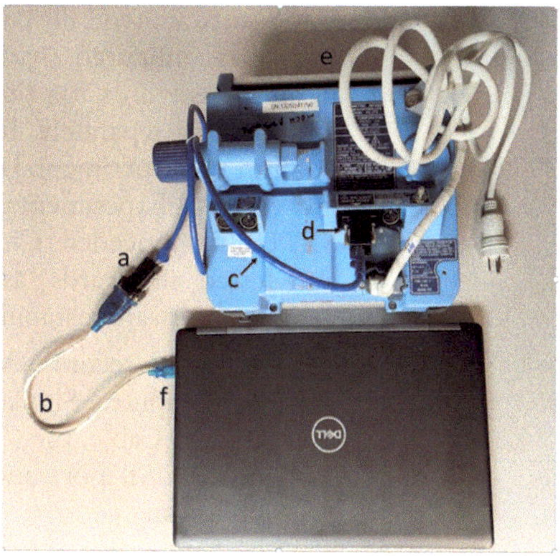

Fig. 3 Computer to Gemini PC-2Tx Cable Connections. (**a**) DB9 to RJ45 Female-to-Female Modular Serial Adapter. (**b**) USB to RS232 DB9 Serial Adapter. (**c**) CAT6 Ethernet Cable. (**d**) DB9 to RJ45 Male-to-Female Modular Serial Adapter. (**e**) Gemini PC-2TX. (**f**) PC

4. Install Secondary Video Monitor
   (a) Alcohol SA experiments require a second monitor for interaction with participants. Connect the monitor to the computer with the video cable. Right click the desktop and select display settings, and for the second monitor, select extend desktop.
   (b) Some investigators want to see the participants' view in real time as the experiment runs, necessitating a third video monitor. After connecting the third monitor, right click the desktop and select display settings, and for the third monitor select Duplicate desktop on 2 and 3.

3.1.2 Initial Gemini PC-2TX Setup

1. Plug the Gemini PC-2TX pump into an outlet and verify that the pump parameters are set for CAIS computer control. The computer control setup procedure needs to be done initially and after service-related resets.
   (a) Press and hold both the "Power On" and "Options/Edit" buttons until the unit powers on to the "Sys Config 1 of 5" menu (hold is approximately 5 s).
   (b) Select choice 4 "C2 Port".
   (c) Select Baud Rate of 9600.
   (d) Enter Serial No 1234.
   (e) Make sure Computer Control Function (#5) is set to "enabled".

3.1.3 Initial CAIS Software Setup

1. CAIS software typically comes loaded onto a secure USB drive and pre-configured with the investigator's experimental design. Passwords for the drive proper and the SETUP file are provided separately. It is recommended that investigators confirm the experimental design parameters in the SETUP file are correct. Experimental adjustments can be made in the SETUP file if needed. Changes should only be done carefully and tested afterwards, as errors may result in experimental errors or software instability.
   (a) Insert the secure CAIS drive into an open USB port on the computer. Similar to the USB to RS232 DB9 Serial Adapter Cable, the CAIS drive should be inserted directly into a USB port and not through a port multiplier or replicator.
   (b) The secure drive may open automatically, but if not, navigate to the drive in Windows Explorer, double click it, and enter the 6-digit drive password.
   (c) Identify the SETUP file (SETUP.mdb), double click and enter setup file password for access. Key parameters are found in the "ESF" table.

3.2 Key Experimental Parameters

3.2.1 Overall Parameters

(a) Experiment number, name, and description identify the protocol.

(b) Breath Limit or Safety Limit: The maximum allowable BrAC (mg/dL) for an experiment. For SA experiments with dynamic rewards, we recommend setting an additional buffer or safety margin to minimize any potential violation of the maximal allowable BrAC secondary to measurement or PBPK modeling error.

(c) ExpDuration or Experimental Duration: The maximal total infusion time (min).

(d) COMPort or Communication Port Variables: Typically set to COM11 for alcohol and, if present, COM12 for alternative rewards.

3.2.2 Investigator--Determined Administration Parameters

(a) Alcohol Clamp: The alcohol clamp is defined by a linear rise to target alcohol exposure which is then maintained. The software, using BrAC measurements as feedback, will adapt the infusion rates as possible to (1) achieve the desired BrAC level at the specified time with the physical pump constraints and (2) maintain that exposure. Participants with large volume of distributions may require infusion rates above the physical pump maximum to achieve specific rates of ascent. BrAC entry results in adjustment of the overall remaining infusion rate trajectory required to achieve the desired clamp exposure.

Target: The desired BrAC (mg/dL) "clamp" level.

Target Time: The designed time (min) at which BrAC will achieve the Target BrAC.

3.2.3 Alcohol Self-Administration Parameters

(a) Reward Properties: The currently available alcohol rewards linearly raise BrAC by a specified amount over a defined duration followed by a controlled, also linear, descent. The software will strive to achieve the reward properties as long as possible given physical pump constraints (participants with large volume of distributions may require infusion rates above the physical pump maximum) and pharmacokinetic state (prolonged or aggressive descents may not be possible later in experiments when participants have reached an equilibrium state). Entry of measured BrACs results in computation of the infusion rates required to achieve the next reward accurately.

Ascending Limb: The desired rate of change in BrAC (mg/dL per min) per reward. For Rate Control paradigms, this corresponds to the maximal available selection. Consequently, and in contrast to the identical reward exposures in FA and PR paradigms, the magnitude of BrAC change per rate control reward is participant dependent.

Reward Duration: The amount of time (min) over which the BrAC will rise.

Descending Slope: The desired rate of reduction in BrAC (mg/dL per min). For FA and PR experimental designs, this descending rate will be maintained as long as possible until the next reward selection. For Rate Control paradigms, this corresponds to the maximal available reduction in BrAC as a reward while the current BrAC is held constant between reward selections.

Alternative Rewards: The availability of alterative rewards is defined in the setup file. The shape and duration of alternative rewards is defined by the infusion rate profile. Currently available options include a fixed sawtooth (linear rise to a predefined level and the linear descent) or "modeled," where the infusion rate profile is consistent with that required for an alcohol reward.

Prime: The use of a priming period during alcohol self-administration is optional. If desired, the duration of the interval, work set demands, and reward properties must be set separately. A Clamp Prime is possible, but its use and interpretation are complicated—It is recommended that authors are consulted before its utilization.

Priming Time: Amount of time (min) for priming interval.

NumPrimingDrinks: Number of priming rewards.

(b) Reward Schedule

Free-Access (FA): As described in the introduction, FA paradigms provide an alcohol reward without cost or work, generally a single button press. This requirement corresponds to a fixed ratio response requirement of 1 (FR1), a commonly used preclinical paradigm. To accomplish the FR1 design, either modify the pre-existing FA experiment type to the experimental needs or create a PR schedule with identical work requirements. The time between successive rewards will be defined by the reward lockout (duration of ascent). The time required to obtain the total number of available rewards should well surpass the experimental duration.

Rate Control: This paradigm functions as an FA paradigm with dynamically adjusted reward properties.

Progressive Ratio (PR): As above, in PR paradigms, each reward self-administration requires completion of either of a specified number of simple button presses or correct Constant Attention Task (CAT) trials which increases as a function of set number. The PR work-schedule is generally study-specific, with reward and work-set parameters interacting to produce a maximal alcohol exposure. The number and demand requirements per work set per reward type are defined in the setup file. The demand per reward is not cumulative. For example, a requirement of 1606 simple button presses for the tenth reward does not include the button presses already made for the ninth reward. The time required to complete the total number of work sets should well surpass the experimental duration.

Constant Attention Task (CAT) Parameters: The CAT task is defined by time intervals over which participant responses are required and the duration of states. An example trial is depicted in Fig. 3.

Dimn and Mimx: minimum and maximum uncertainty interval durations, respectively (seconds). 1.5 and 3.5 s in the example in Fig. 4.

Drs: Response window duration (msec). Common initial setting for DrS in adults is 300 ms.

dDr, iDr, and Drmx: incremental decrease, increase, and maximal duration of the response window, respectively (ms). 50% success rates goals require equal dDr and iDr, commonly 25 ms.

Dfb, Fb, and Afb: delay until Feedback, duration of feedback, and the amount of time after feedback until the next CAT trial may be initiated (ms). Dfb and Fb durations of 0.5 s are displayed in Fig. 3.

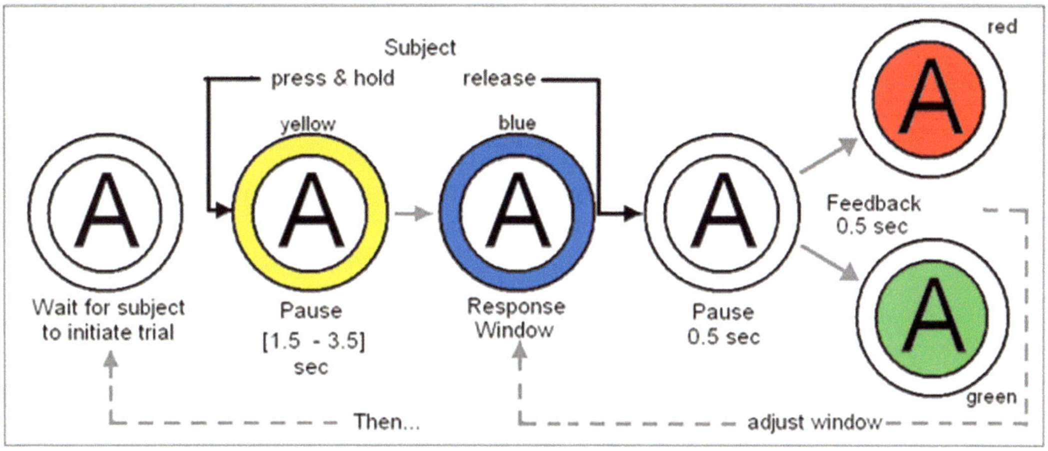

Fig. 4 A Constant Attention Task (CAT) Trial. Participants start a trial by pressing and holding down a button. The outer ring turns from white to yellow signifying the beginning of an anticipatory interval of unpredictable duration (uniformly distributed over 1.5–3.5 s). The outer ring then changes to blue. To count as a success, the participant *releases* the button within the blue response window; i.e., before the outer ring returns to white. Feedback is provided about the trial outcome (green for correct, red for incorrect). Response windows adjust so that success rate remains near 50% regardless of any effects of alcohol or fatigue

Simple Button Press: No specific parameters beyond the PR response requirements are required.

3.3 Participant Recruitment and Consent

1. Pre-screen participants via short telephone interview to determine if they meet basic eligibility criteria. The answers to these questions should not be stored with any identifying information as they have not yet consented to the study and are only being asked the questions for eligibility and safety. All study-related procedures should be conducted following appropriate institutional ethics review board approval and consistent with applicable human subjects research regulations and policies of the institution. It is also recommended that the investigation follows the NIAAA Council guidelines for administration of alcohol for research in humans [102].

   (a) Participant baseline drinking pattern is an important eligibility and safety criterion. Generally, participants should report a drinking pattern that suggests they will be able to safely tolerate the BrAC exposure when they are infused during participation. Consequently, the requirements should change relative to the experimental design, including the population being recruited, for optimal participant safety. For example, it would not be advisable to clamp light alcohol drinkers at 80 mg/dL (binge drinking definition), if that concentration is beyond what they normally experience. The phone screen should include questions about normal drinking, including how many drinks the participant consumes each day of an average week.

(b) Depending on the nature of the study and local requirements, it may also be important to exclude any participant who reports a treatment seeking status or who endorses withdrawal symptoms from alcohol that would suggest they cannot abstain from drinking alcohol long enough to arrive with zero BrAC. Participants are typically instructed to stop drinking the day before the experiment early enough to ensure there is no alcohol detected at experimental check-in time. The recommendation for an infusion duration will vary with the population being studied and other experimental design considerations (such as when the experiment will be run).

(c) Participants should also be screened for any clinically significant medical problems or conditions that might impact safety during participation or data quality. These conditions and diseases include pregnancy, high blood pressure, and liver or pancreas conditions such as Hepatitis C, cirrhosis, or pancreatitis. Note: Aside from pregnancy, these are not always immediately disqualifying factors. For example, properly managed or monitored high blood pressure may be acceptable with monitoring. Other conditions may require consultation with the participant's physician to ensure safe participation.

(d) There are several recreational drugs and medications that should not be mixed with alcohol and so should be asked about during the initial phone screen. A participant who endorses regular illicit drug or other substance use that would put them at risk during an alcohol experiment should be disqualified from the study. Marijuana use is typically not a disqualifying factor, but in any case, the participant should be told not to use any drugs the day of the study to avoid any reactions with alcohol. The phone screen should ask about any prescription or over-the-counter (OTC) medication that the participant has taken in the last 30 days. If there are any known alcohol-medication interactions, the principal investigator (PI) or medical director should be consulted as appropriate.

2. Once participants pass the initial phone screen, schedule them in for a more in-depth initial screening visit.

   (a) The participant should consent and sign a consent form before any screening visit procedures are conducted. Standard procedures included in this visit are medical history and brief physical examination, emergency contact information collection, self-report measures evaluating levels of depression or anxiety that might indicate increased participant risk, a baseline withdrawal questionnaire such as the CIWA, and a urine screen for pregnancy (if female) and illicit drugs or prescriptions that interact with alcohol. The goal of the brief physical exam is primarily to identify potential problems with peripheral IV access.

(b) Optional procedures that are sometimes included in the screening visit might include any baseline self-report or interview measures needed for study procedures or for screening based on other factors in the study procedures (neuroimaging, electroencephalography, fMRI, post-traumatic stress disorder, etc.), history of flushing response, family history information including medical history and difficulties with alcohol or drugs, or an electrocardiogram. Depending on the experimental design and participant population, blood tests for liver function as well as others should also be considered.

(c) Additionally, participants should have their full recent drinking history assessed using the Timeline Followback (TLFB) of duration determined by the study design.

3. Review each participant's screening data to determine eligibility for the study.

(a) Exclude participants if they have any clinically significant medical or other problems determined by the measures and/or PI to be unsafe for participation, if they use prescription or OTC medication known to interact with alcohol that cannot be taken the day of the study as determined by the PI, if they are currently seeking treatment for alcohol use disorder, or if females state an intention to become pregnant. In any case, urine testing for pregnancy should be performed for females on the day scheduled for the experiment.

3.4 Session-Related Methods

3.4.1 Baseline Measures and Setup

(a) Participant arrives on the day of testing after abstaining from alcohol long enough to ensure a zero BrAC at the start of the session.

(b) Perform a breathalyzer test to ensure a zero BrAC.

(c) Take vitals of interest such as temperature, respiration rate, blood pressure, and heart rate. Confirm age and biological sex at birth. Take height (cm) and weight (kg).

(d) Collect a sample for a urine drug screen for all participants as determined by the specific protocol. Run a urine beta-hCG pregnancy test for females.

(e) Depending upon the session arrival time, provide participant a meal. A standardized meal is recommended to minimize the effects of meal timing and composition on gastrointestinal blood flow and secondary effects on alcohol metabolism.

(f) Complete a brief medical history to determine any recent hospital visits, illnesses, new medications since their screening visit for any changes that may significantly impact their safety during the study and quality of the data collected. Consider administering a recent drinking history questionnaire to assess any changes in alcohol consumption since their screening visit.

1. Alert nursing and research staff to any clinically significant medical findings.

(g) Insert a 20 ga IV catheter into a vein in the antecubital fossa of the non-dominant arm for the alcohol infusion. Secure the positioning, heparinize, and cap the catheter.

1. Dominant hand will be used for answering questionnaires during session.

(h) Confirm delivery of alcohol infusate, including participant name, age, sex, and date of expiration.

1. Optional: Measurement of Alcohol Infusate Concentration. The standard alcohol infusate concentration should be 6% ethanol (v/V), using at least half-normal saline as the vehicle. It is recommended to test the infusate alcohol concentration with a refractometer as follows.
2. Cover the lens at the base of the basin of the refractometer with distilled water, plus a little extra (you don't need to fill it all the way).
3. Turn on the refractometer, and once it's ready, press the "Zero" button.
4. Wipe the lens with chem wipes until completely dry.
5. Place about 5–6 drops of the alcohol infusate in the basin. Again, use enough infusate to cover lens, plus a little extra. The infusate may be obtained by running the infusion pump at a rate of 400 mL/h.
6. Hit the "Start" button and record the reading. It should be generally between 2.5 to 2.9 for alcohol infusate concentrations of approximately 6% alcohol by volume.
7. Wipe the lens with chem wipes and repeat **steps 4–5** to complete two more readings (three total per alcohol infusate bag or bottle).
8. Zero the refractometer again per **step 2**, then repeat **step 3–6** on any other supplies of alcohol infusate.
9. Input all six readings into the refractometer calculator (provided with the CAIS secure USB drive) to obtain the alcohol infusate concentration. Depending on the calculator version, settings for the refractometer and alcohol

infusate vehicle (for example, 1/2 normal saline) may be required. If only one supply of alcohol infusate is present, duplicate the three readings to complete the missing entries. Alternatively, the formula for calculating the alcohol concentrations is

Half normal saline [(mean of refractometer readings—0.5579)/0.34702] = Percent alcohol

Ringers Lactate ([mean of refractometer readings – 0.9357]/0.34915) = Percent alcohol

(i) Instruct participant to take a bathroom break before starting the infusion to minimize experimental disruption.

1. The bathroom should be easily accessible to the participant during the infusion. If the bathroom is in a different room, the participant will need to be disconnected from the IV fluid lines. If the bathroom is close enough that the participant can stay connected while using the restroom, an extension will be needed for the IV line.

(j) Consider obtaining another blood pressure and breathalyzer measurement for safety.

(k) Connect the infusion tubing to the infusate supply bags or bottles and purge the tubing of any air bubbles. If possible, choose tubing lengths between the participant's arm to be infused and IV pump based on proximity to the bathroom so that the participant may use the restroom while remaining connected.

1. Please see Notes – Connecting Two Pump Channels to a Single Source and Insufficient Alcohol Infusate for an alternative and troubleshooting connection strategy.

(l) Infusion Pump Setup and Computer Connection.

1. Switch on the pump by pressing the gray "POWER ON" button on the upper right side.
2. Press the "VOLUME INFUSED" button on the left side. If the numbers do not read zero, press the "CLEAR" button.
3. Press the "OPTIONS/EDIT" button.
4. Press the number 4 to select the computer control option.
5. Use the arrow buttons on the top row of the pump to select "YES" and press "ENTER." "Computer Control" should now be flashing on the top of the infusion pump screen.
6. Please see Notes – Infusion Pump Communication Failure for Troubleshooting Procedure.

3.4.2 CAIS Laptop Setup

1. Turn off Wi-Fi before running an experiment.
2. If needed for experimental paradigm, connect the participant response buttons or other interface buttons. For alcohol self-administration experiments, the buttons will determine participant selection of alcohol rewards. Connect response button into the USB portal closest to the participant.
   (a) Participant button must be connected prior to opening up CAIS software.
3. Initialize CAIS Software
   (a) Insert the CAIS secure USB drive.
   (b) Select drive to enter password.
   (c) Once password has been accepted, open the drive titled "SecuDrive".
   (d) Double click the CAIS software icon and the program will initiate.
   (e) Create Data File.
      1. Click "Select File" and "New Session".
      2. Create filename using desired participant number and experiment file name format.
      3. Save the data to the desktop or local folder—Do not save participant data directly to the USB secure drive to prevent read/write conflicts, which can result in data loss. During the session, this computer will not have an internet connection, but files can be transferred after completion of the experiment.
   (f) Select the experiment paradigm. The most common alcohol infusion experiment paradigms fall into three categories: Progressive Ratio, Free Access, and Clamp.
   (g) Enter Participant Morphometrics (*see* Fig. 5).
      1. Enter the participant number.
      2. Choose Transform "Version Three" for participants who are younger than 30 years of age, "Standard" for those 30 years of age and above.
      3. Enter biological sex at birth and age.
      4. Enter height (cm) and weight (kg) measurements from arrival.
      5. Click "submit" to enter details into CAIS.
   (h) The CAIS graph panel will now appear on the screen with the participant display superimposed (*see* Fig. 6). Move the participant display window over to the appropriate monitor by clicking and dragging across the screen. Enlarge the participant display window to occupy the entirety of the participant monitor.

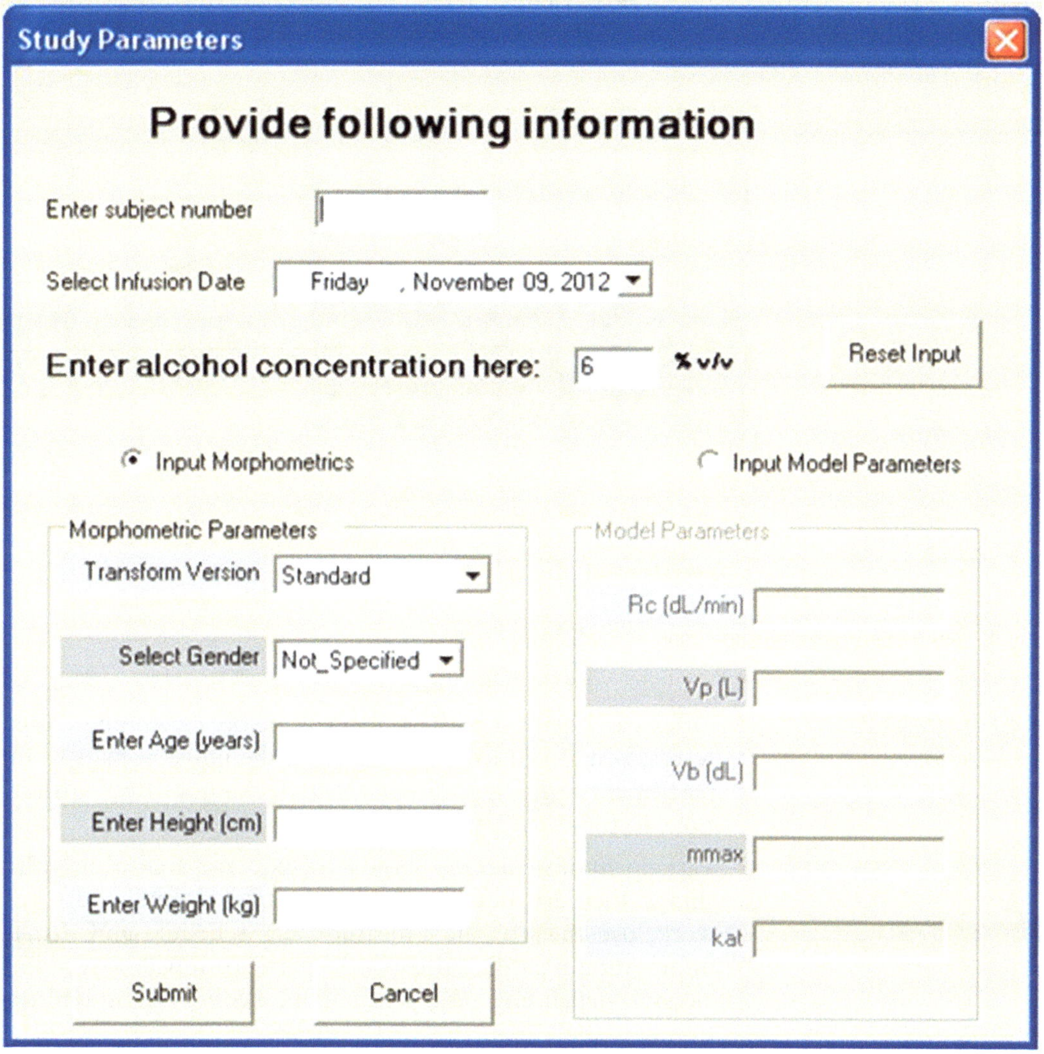

Fig. 5 Participant morphometric entry. The subject characteristics will determine the initial set of physiologically based pharmacokinetic model parameters. This parameter set, in combination with the desired exposure trajectory and eventually BrAC entry, will determine the infusion rate time course

(i) If the paradigm requires self-administration buttons, their electronic address will be "mapped" after entering participants details. A small dialog box will pop up with sections corresponding to each reward. The user presses each button until the blank space in the dialog boxes is populated, and then press "Ok".

1. Buttons can be remapped at any point in the experiment by navigating to the View menu, then clicking "Map Drink Buttons".
2. Please see Notes – CAIS Button Troubleshooting for additional debugging procedures.

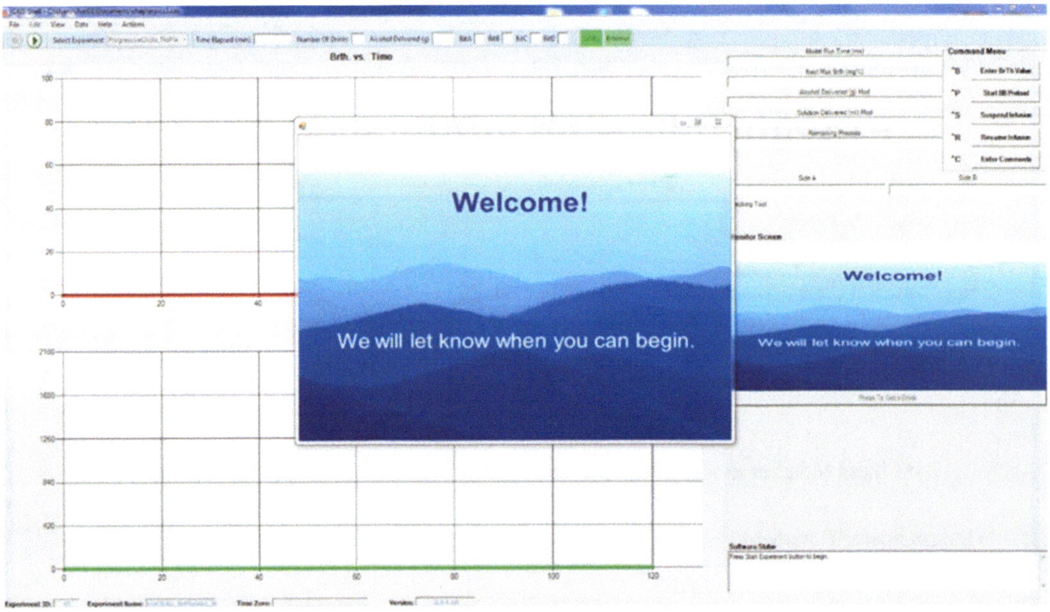

Fig. 6 CAIS graph panel with superimposed participant display

3.4.3 CAIS Session Initialization

1. Offer participant a bathroom break before beginning. Afterwards, connect infusion tubing from the infusion bags to the participant using appropriate sterile technique. Read instructions to the participant.
2. Participant instructions should include the following: what things they are not allowed to do (for example, read, do work, use their phone), measurement schedule and details, information regarding the way the participant will be receiving alcohol, depending on the paradigm, alcohol self-administration basics.
3. Be sure to collect any baseline measures from the participant before clicking start on the CAIS program to ensure you have these measures at a 0.000% (0.0 mg/dL) BrAC.
4. On the CAIS laptop, click the start/run ("Play") button at the top of the screen. The experiment will begin and the participant's screen will reflect the first steps, depending on the paradigm.
   (a) Please see Notes – IV Line Occlusion for Troubleshooting procedures.
5. Breathalyzer readings are typically collected at least every 15 min until the end of the infusion. BrAC measures are entered immediately into CAIS so that the system can adapt to modeling errors. Subjective responses or other dependent measures can also be collected at these same timepoints.

(a) Enter the breathalyzer reading into the CAIS software by pressing ctrl+B or selecting "Enter BrTh Value" and typing in the value (i.e., 0.030).

(b) Breathalyzer reading frequency should also be adjusted based on modeling accuracy. For example, if the breathalyzer readings are much higher or lower than CAIS predicted, the investigator should take additional measurements sooner and more frequently until the model and measurements converge.

(c) Please see Notes – Incorrect BrAC Entry/Cancelled BrAC Entry for Troubleshooting procedures.

6. Bathroom breaks can be administered in a few different ways, depending on experiment setting and procedures. Care must be taken during bathroom breaks to help ensure the participant can safely move to and from the bathroom. Investigators should consider basic sobriety checks or transport assistance procedures depending on the experimental design and population of interest.

(a) If the bathroom is in close proximity with privacy for the participant to stay connected to the IV line, the participant can use the bathroom without any action needed in CAIS. However, it is important to help the participant maintain safety by ensuring equipment does not become a trip hazard. Record the time the bathroom break started and ended.

(b) If the bathroom is too far for the participant to stay connected to the IV line, a bathroom break process will need to be initiated within the CAIS software. This can be done by pressing the "Start BB Preload" button in CAIS, which will initiate a "pre-load" period in the infusion. This two-minute infusion rate provides a small boost in alcohol levels to counter the decrease that occurs when the infusion is paused and the participant is disconnected from the pump for the bathroom break. The participant's screen will change during their bathroom break, and they will be instructed not to press any buttons or request rewards during the pre-load.

(c) Sometimes, a pre-load is not required or preferred for a bathroom break. This can be due to the urgency of the participant's need for a bathroom break, in which case the infusion can be suspended in CAIS without pressing the bathroom break button. An immediate suspension might also be preferred during a free access experiment, in which it is not important to maintain the participant at a specific BrAC level.

1. Breathalyzer measurement should be delayed for at least 5–10 min after any bathroom break to account for a) the pre- and post-load or suspension procedure and b) alteration in the pharmacokinetics of alcohol through changes in heart rate, availability of rapidly equilibrating volumes, etc. Using breathalyzer readings obtained during this time will introduce these transient changes into the infusion rate calculations.

7. Note that the infusion ends either by pressing the stop button or after the protocol specific duration. CAIS will create a .CAS file of all time-stamped data relevant to the experiment, including participant number, dates, morphometric data, PBPK parameters, BrAC trajectory, infusion rate profile, BrAC measurements, button push history, alcohol used, bathroom breaks, and any technician comments entered during the session. The .CAS file is write-protected, but can be used to replay the session at any future date. The .CAS file is exportable to Excel for analyses.

3.4.4 Paradigm-Specific Procedures

1. *Alcohol Clamp Procedures.* The clamp can serve as a stand-alone experiment type or as a priming exposure, depending on the protocol as discussed above.

   A. The participant should see a screen indicating that the clamp has begun and, depending on the experimental design, inform them of the amount of time until their BrAC reaches the target level.

      1. During this period, it is typically okay for the participant to relax, and depending on investigator preference, they could be allowed to use their phone to listen to music or read. In general, talking on the phone is discouraged during the infusion session.

   B. During the clamp, CAIS will begin directing the pumps to infuse alcohol to the participant at a rate based on the details entered before the experiment (sex, height, weight, etc.) Once the participant's BrAC is predicted to be at about 0.020%, take a breathalyzer measurement and enter the value into CAIS to correct for any errors in the CAIS model.

   C. Depending on whether the BrAC measurement is high or low in comparison to the CAIS-predicted value, CAIS will adjust the infusion rate time course to return to the desired exposure. These changes will be reflected with updates in the infusion rate graph in the bottom half of the CAIS screen.

   D. Breathalyzer readings should be taken periodically during the ascending limb to maximize infusion rate calculation accuracy and thus precision in reaching the target at the desired time. Breathalyzer readings begin at approximately 20 mg/dL (secondary to breathalyzer device sensitivity characteristics at low concentrations) and then with every approximate 10 mg/dL BrAC increase until the target is reached.

E. Once the target is estimated to have been achieved, a breathalyzer reading should be collected. The transition from ascending limb to maintenance of target concentration commonly requires some BrAC measurement-based infusion rate adjustments over the first several minutes.

F. Subsequent breathalyzer readings are taken periodically throughout the experiment and as necessary, given the dependent measure battery timing, bathroom breaks, etc.

G. Depending on the experimental procedures, self-report measures are typically administered at the clamp level to determine any differences in these measures compared to baseline (pre-experiment).

H. The investigator should check in with the participant throughout the duration of the clamp period. If an IV-line occlusion were to occur (*see* Notes for additional details), it is most detectable in the first few minutes of the infusion. Signs of an occlusion include feelings of burning or pain, bruising, or toughness in the arm being infused. The investigator should periodically be asking the participant and physically checking the infusion site for these symptoms.

2. Alcohol Self-Administration Procedures

   1. Optional Priming Interval

      (a) Instruct the participant to press the button as soon as the monitor prompts them to for their priming doses. Remind the participant to begin pressing the button each time the screen prompt appears so that they reach their expected BrAC consistent with the experimental design.

      (b) Obtain a breathalyzer reading at the peak prime exposure if greater than 20 mg/dL (given limitation in breathalyzer sensitivity at low BrAC concentrations).

      (c) The algorithm will attempt to adjust the infusion rate profile for subsequent rewards using all breathalyzer measurement information to achieve the desired reward exposure parameters each time a BrAC is entered.

         1. Collect prime-related dependent measures as applicable.

         2. Allow participant to rest for the remainder of the priming interval. CAIS will continue to deliver the

infusate with a rate profile with a goal of the priming reward descent rate. No further alcohol or alternative rewards will be delivered until the self-administration interval begins.

2. Alcohol Self-Administration Interval

   A. Participant instructions will vary based on the experimental design, including dependent measure battery. Generally, the user will instruct on self-administration requirements, with limited direction provided by the software. Generally, a screen appears instructing the participant that the "Bar is open" and they can receive an alcohol reward. When a reward is selected or earned, the pre-determined alcohol infusion rate profile to achieve the incremental exposure trajectory is delivered to the participant, while the screen says that the "Bar is closed" until, as defined by the experimental protocol, the incremental ascending limb is complete, or the lockout period has elapsed.

   B. During reward delivery, it is recommended that the investigator collect dependent measures or communicate with the participant. When, as defined by the experimental protocol, the incremental ascending limb is complete, or the lockout period has elapsed the alcohol infusion is complete, the screen returns to the "Bar is open" screen, and the participant is free to continue self-administration.

3. Throughout the experiment, it is essential that the user collect and enter breathalyzer readings into the CAIS software. BrAC measurements should be taken every 20 mg/dL of ascent and periodically throughout the experiment. It is recommended that readings be obtained near incremental reward peak BrAC while the "Bar is closed" to minimize disruption of the self-administration interval.

   (a) As in the priming interval, the alcohol infusion rate profile for subsequent rewards is recalculated using all available information.

   (b) Also, as in the priming interval, changes in physical activity will alter the pharmacokinetics of alcohol. Recommend waiting at least 5 min *after* these events before a breathalyzer reading is entered.

4. The participant screen will also indicate the "Bar is closed" when the delivery of an alcohol reward would result in a violation of the predefined breath alcohol limit. The determination is based on the modeled BrAC trajectory over the entire experiment (in contrast to the most recent BrAC measurement).

(a) For dynamically calculated alcohol rewards, the maximum increase in rate of BrAC change will be constrained such that delivery will reach the safety limit and measurement buffer. If re-computation of the BrAC trajectory results in an estimate above the buffer or safety limit, the BrAC will be forced to reduce until the safety condition is no longer violated.

5. Non-SA-Dependent Measure Collection: The timing of other dependent measure collection is protocol dependent, but recommends timing during "Bar is closed" intervals to limit SA disruption.

3.4.5 End of Session

1. Click the Data tab on the CAIS software and select "Export to Excel".
2. Remove the IV catheter using sterile technique.
3. Record the total volume of infusate used on the flowsheet.
   A. Press the "VOLUME INFUSED" button on the left side of the pump to see the values and record the total volume of ethanol administered, which may be required for clinical documentation.
4. Collect a breathalyzer reading every 15–30 min until BrAC is below 0.020 g/dL (20 mg%), or until the study-specific discharge criteria are met.

3.4.6 Outcome Analysis

The outcome measure list below is not meant to be exhaustive, but rather an inventory of common metrics.

1. Clamp Exposure-Based Outcome Measures
   (a) Common clamp-based measures include average exposure during the dependent measure batteries and average time to achieve the target concentration.
   (b) Other clamp-based measures include steady-state alcohol infusion rate or descending limb rate of changes (commonly used for measurement of alcohol elimination rate).
2. Alcohol SA Outcome Measures
   (a) General measures included peak BrAC, average BrAC, cumulative BrAC (integral of BrAC Curve), total number of rewards earned, and total ethanol consumed.
   (b) These measures generally do not include the priming portion of the session.
   (c) CAIS FA and progressive ratio measures of interest include time to 80 mg/dL (or other targets), cumulative time above a specific target.
   (d) CAIS PR measures of interest will vary across the protocol, but may include cumulative work (total number of

button presses) by reward type, breakpoint (the amount of work in the final completed work set), seeking time or fraction (the amount of time spent working for a specific reward), average rate of button press, and for the CAT: accuracy, response window, and the number of impulsive button presses (button release before the trigger).

4 Notes

4.1 IV Line Occlusion

A blockage of the free flow of liquid through an IV line is called an occlusion. The most common cause of an occlusion happens when someone or something blocks the IV line before it reaches the participant, such as a heavy object being placed on the IV line. It may also be that the participant moves the line or catheter into a position that kinks the line and causes the occlusion. Clinically designed intravenous infusion pumps attempt to automatically detect this event through comparison of an indirect pressure measurement to internal settings. Occlusions can be detected by the operator by (a) physical examination of the antecubital space—fluid not entering the vein will cause some swelling near the IV catheter or (b) a consistent mismatch between the infusion rate and BrAC measurements such as stable or decreasing measurements despite a rate that should increase them. The Gemini PC-2Tx pumps sound an occlusion alarm when an occlusion is detected and cease delivery until the problem is resolved. For cases like these, the solution is to remove the blockage to allow the free flow of fluid. The pump will reset itself within 30 s and the experiment can continue. A less common cause of occlusion occurs when there is a blockage within the IV catheter or the vein itself. After the catheter is set, it can sometimes move into a position that impedes flow, such as if the opening is resting next to a vein valve, and the catheter may need to be reset. Typically, a nurse or phlebotomist can flush the line out or reset the line to remove the blockage. This procedure may take a few minutes and it is usually best to pause the infusion until the occlusion is removed.

4.2 Incorrect BrAC Entry/Cancelled BrAC Entry

CAIS is not able to "undo" incorrect breath reading entries, but several strategies exist to mitigate the impact of these readings and minimize the associated infusion rate profile adjustment. The impact of a single erroneous error measurement will be larger at the beginning of an experiment (as fewer total breathalyzer readings have been entered) and for clamp experiments (which use only the last several measurement values to force the participant exposure closer to the target). One compensatory strategy for clamp experiments is to immediately enter a value with an offset based upon the original error. For example, if the participant's BrAC is 80 mg/dL, and 70 mg/dL is entered, the investigator could immediately enter a breath of 90 mg/dL to offset the original error and minimize overcorrection in the updated infusion rate calculation. Another commonly used strategy is to collect and enter additional breathalyzer readings after the error until the rate calculation appears to stabilize. This additional breathalyzer measurement approach is also used when there is concern about a specific measurement, for example, if a breathalyzer reading is markedly different from the expected estimate in an experiment that was previously stable, if the staff perceives that a deep breath sample was not provided by the participant, etc. However, if a breath entry process is initiated but then cancelled, the next measurement must be entered twice in quick succession to prevent an erroneous rate calculation.

4.3 Connecting Two Pump Channels to a Single Source and Insufficient Alcohol Infusate

The procedures below can be used to supply two infusion pump channels with a single alcohol infusate supply. The process can also be used if a new supply must be added or one supply of two will empty before the other secondary to idiosyncratic filling differences. (It is generally recommended that investigators simulate protocols calling for very long infusions or those experiments testing participants with large volumes of distribution to ensure an adequate alcohol infusate supply is available.)

1. To create a bridge between two sources, run a Baxa extension set male LL to male LL, ref 87 connector between the two connection ports above the pump (closer to the alcohol infusate) using sterile technique. Scrub the port connection with an alcohol swab, attach the male-to-male tubing, and prime the line to remove air. Then, clean the other line connection with the alcohol swab and attach to the second line. One alcohol infusate source will now supply both infusion pump channels.
2. To add another source, using sterile technique, while keeping the line drip chamber upright, quickly remove the spike from a depleted alcohol infusate source *without letting it touch anything*. Insert the spike quickly into the new alcohol infusate source without letting the drip chamber empty. The drip chamber must stay upright during this process to avoid introduction of air into the lines. If the infusion rate is low enough to prevent emptying the drip chamber, the spike may be wiped with an alcohol swab, if necessary.
   (a) Optional Refractometer Usage: Once the additional supply of alcohol infusate is obtained and the original sources are almost empty, and before the step above, use sterile technique to carefully draw 5–10 mL of the new infusate. Test the alcohol concentration with the refractometer as previously described. The new measured concentration cannot be entered once a CAIS session has been initiated,

but it should not be used if too high. After obtaining alcohol infusate for concentration testing, clamp the line port on the bag using the hemostats and proceed as above.

(b) Once the new supply of alcohol infusate has been connected, use hemostats to clamp off the other depleted bag beneath the drip chamber. Now, with the bridge in place above the level of the infusion pumps, the new alcohol infusate supply will flow to both pump channels.

4.4 Infusion Pump Communication Failure

1. Initial Infusion Pump Communication Failure: Sometimes, the computer control option is not available on the pump or computer control is available, but when the user starts the experiment, the software warns that no pumps are connected. These conditions are most caused by loose physical connections between the computer and the infusion pumps. The problem is resolved by correcting any loose or broken connections and then restarting CAIS.
   (a) Less common causes include incorrect device or software settings. After pump servicing (for example, to replace a battery), default settings are restored. In that case, follow initial pump programming instructions and restart the experiment. Finally, the pumps will not connect to the computer if the setup file port designations do not match the computer settings—see initial setup instructions for details.
2. Mid-Experiment Infusion Pump Communication Failure: On extremely rare occasions, the CAIS software will stop communicating with the infusion pumps or freeze in the middle of an experiment secondary to a software conflict or other program incompatibility (see Initial setup details for instructions on how to minimize these events). In this event, it may be necessary to stop and restart the experiment, or to manually adjust the flow rates on the pumps to achieve the desired BrAC trajectory. An alcohol clamp can continue by manually adjusting the flow rates to achieve the desired BrAC using the most recent estimate as a guide. Breaths should be measured frequently as described in previous sections, and flow rates adjusted accordingly to maintain the desired target BrAC. Self-administration experiments typically have accompanying graphic stimuli and instructions that cannot be restarted and usually must be restarted.

4.5 CAIS Button Troubleshooting

1. Do not stop the session if the response button loses connection with the CAIS software and no longer responds to a press.
2. Remove button from the port and place it in a different port on the computer. The original port is no longer recognized by the software.

3. Click "View" on the CAIS software screen and scroll to the bottom and select "Map Drink Buttons". A pop-up window will show up on your screen.
   (a) Press the drink button(s). The new address information will enter the pop-up window.
   (b) Select "OK" on the pop-up window on the computer screen.
4. The experiment can now proceed as before.

Acknowledgments

We would like to gratefully acknowledge NIAAA funding of the Indiana Alcohol Research Center (MHP, AEK, SOC; P60 AA07611) and R01 AA027236 (MHP) support for the Computer-assisted Alcohol Infusion System-based research studies. Studies at IU also benefit from the support of the Indiana Clinical and Translational Institute Clinical Research Center (prior M01 RR 750, now UL1TR001108) support and, for the preparation of 6% alcohol v/v infusate, the IU research pharmacy. VAR and BLS are supported by NIAAA Division of Intramural Clinical and Biological Research (Z1A AA000466). The programming expertise of Victor Vitvitskiy remains vital to these projects and is most sincerely appreciated.

Invitations: Investigator interest in adapting the capabilities of the Computer-assisted Alcohol Infusion System to their own research is welcome and those interested should send an email to mplaweck@iupui.edu.

References

1. Li TK et al (1979) Progress toward a voluntary oral consumption model of alcoholism. Drug Alcohol Depend 4(1–2):45–60
2. Gordon TL, Meehan SM, Schechter MD (1993) P and NP rats respond differently to the discriminative stimulus effects of nicotine. Pharmacol Biochem Behav 45(2): 305–308
3. Ritz MC et al (1994) Operant ethanol-reinforced behavior in P, NP, HAD, and LAD rats bred for high versus low ethanol preference. Alcohol Clin Exp Res 18(6): 1406–1415
4. Samson HH et al (1998) Comparison of alcohol-preferring and nonpreferring selectively bred rat lines. I. Ethanol initiation and limited access operant self-administration. Alcohol Clin Exp Res 22(9):2133–2146
5. June HL et al (2004) Central opioid receptors differentially regulate the nalmefene-induced suppression of ethanol- and saccharin-reinforced behaviors in alcohol-preferring (P) rats. Neuropsychopharmacology 29(2): 285–299
6. Eiler WJ 2nd et al (2007) Responding for brain stimulation reward in the bed nucleus of the stria terminalis in alcohol-preferring rats following alcohol and amphetamine pretreatments. Synapse 61(11):912–924
7. Ramchandani VA et al (2009) Intravenous ethanol infusions can mimic the time course of breath alcohol concentrations following oral alcohol administration in healthy volunteers. Alcohol Clin Exp Res 33(5):938–944
8. Kosobud AE et al (2015) Adaptation of subjective responses to alcohol is affected by an interaction of GABRA2 genotype and recent

drinking. Alcohol Clin Exp Res 39(7): 1148–1157
9. Plawecki MH et al (2018) Alcohol affects the P3 component of an adaptive stop signal task ERP. Alcohol 70:1–10
10. Gomez R et al (2012) Intravenous ethanol infusion decreases human cortical gamma-aminobutyric acid and N-acetylaspartate as measured with proton magnetic resonance spectroscopy at 4 tesla. Biol Psychiatry 71(3):239–246
11. Sutker PB et al (1987) Acute alcohol intoxication: sex comparisons on pharmacokinetic and mood measures. Alcohol Clin Exp Res 11(6):507–512
12. Mishra L et al (1989) More rapid elimination of alcohol in women as compared to their male siblings. Alcohol Clin Exp Res 13(6): 752–754
13. Frezza M et al (1990) High blood alcohol levels in women. The role of decreased gastric alcohol dehydrogenase activity and first-pass metabolism. N Engl J Med 322(2):95–99
14. Kwo PY et al (1998) Gender differences in alcohol metabolism: relationship to liver volume and effect of adjusting for body mass. Gastroenterology 115(6):1552–1557
15. Li TK et al (2000) Variation in ethanol pharmacokinetics and perceived gender and ethnic differences in alcohol elimination. Alcohol Clin Exp Res 24(4):415–416
16. Ramchandani VA, Kwo PY, Li TK (2001) Effect of food and food composition on alcohol elimination rates in healthy men and women. J Clin Pharmacol 41(12):1345–1350
17. McCaul ME et al (2018) A paradigm for examining stress effects on alcohol-motivated behaviors in participants with alcohol use disorder. Addict Biol 23(2):836–845
18. O'Malley SS et al (2002) Naltrexone decreases craving and alcohol self-administration in alcohol-dependent subjects and activates the hypothalamo-pituitary-adrenocortical axis. Psychopharmacology 160(1): 19–29
19. Drobes DJ et al (2003) A clinical laboratory paradigm for evaluating medication effects on alcohol consumption: naltrexone and nalmefene. Neuropsychopharmacology 28(4): 755–764
20. McKee SA et al (2009) Varenicline reduces alcohol self-administration in heavy-drinking smokers. Biol Psychiatry 66(2):185–190
21. Chapman ER, Williams PT Jr (1951) Intravenous alcohol as an obstetrical analgesia. Am J Obstet Gynecol 61(3):676–679
22. Korsten MA et al (1975) High blood acetaldehyde levels after ethanol administration. Difference between alcoholic and nonalcoholic subjects. N Engl J Med 292(8):386–389
23. Gibbens GL, Chard T (1976) Observations on maternal oxytocin release during human labor and the effect of intravenous alcohol administration. Am J Obstet Gynecol 126(2):243–246
24. Nishimura M, Hasumura Y, Takeuchi J (1980) Effect of an intravenous infusion of ethanol on serum enzymes and lipids in patients with alcoholic liver disease. Gastroenterology 78(4):691–695
25. Hansbrough JF et al (1984) Administration of intravenous alcohol for prevention of withdrawal in alcoholic burn patients. Am J Surg 148(2):266–269
26. Jones AW, Norberg A, Hahn RG (1997) Concentration-time profiles of ethanol in arterial and venous blood and end-expired breath during and after intravenous infusion. J Forensic Sci 42(6):1088–1094
27. Ray LA, Hutchison KE (2004) A polymorphism of the mu-opioid receptor gene (OPRM1) and sensitivity to the effects of alcohol in humans. Alcohol Clin Exp Res 28(12):1789–1795
28. Ray LA et al (2007) Pharmacological effects of naltrexone and intravenous alcohol on craving for cigarettes among light smokers: a pilot study. Psychopharmacology 193(4): 449–456
29. Aalto S et al (2015) Intravenous ethanol increases dopamine release in the ventral striatum in humans: PET study using bolus-plus-infusion administration of [(11)C]raclopride. J Cereb Blood Flow Metab 35(3):424–431
30. Westman JG, Bujarski S, Ray LA (2017) Impulsivity moderates subjective responses to alcohol in alcohol-dependent individuals. Alcohol Alcohol 52(2):249–255
31. Ramchandani VA et al (1999) A physiologically-based pharmacokinetic (PBPK) model for alcohol facilitates rapid BrAC clamping. Alcohol Clin Exp Res 23(4): 617–623
32. O'Connor S, Ramchandani VA, Li TK (2000) PBPK modeling as a basis for achieving a steady BrAC of 60 +/− 5 mg% within ten minutes. Alcohol Clin Exp Res 24(4): 426–427
33. Plawecki MH et al (2008) Physiologically based pharmacokinetic (PBPK) models for ethanol. IEEE Trans Biomed Eng 55(12): 2691–2700

34. Plawecki MH et al (2004) Estimation of ethanol infusion profile to produce specified BrAC time course using physiologically-based pharmacokinetic (PBPK) models. Conf Proc IEEE Eng Med Biol Soc 2006:778–781
35. Zimmermann US et al (2008) Development and pilot validation of computer-assisted self-infusion of ethanol (CASE): a new method to study alcohol self-administration in humans. Alcohol Clin Exp Res 32(7):1321–1328
36. Zimmermann US et al (2009) Offspring of parents with an alcohol use disorder prefer higher levels of brain alcohol exposure in experiments involving computer-assisted self-infusion of ethanol (CASE). Psychopharmacology 202(4):689–697
37. Plawecki MH et al (2013) Voluntary intravenous self-administration of alcohol detects an interaction between GABAergic manipulation and GABRG1 polymorphism genotype: a pilot study. Alcohol Clin Exp Res 37 Suppl 1(01):E152–E160
38. Plawecki MH et al (2019) Comparison of subjective responses to oral and intravenous alcohol administration under similar systemic exposures. Alcohol Clin Exp Res 43(4):597–606
39. Cyders MA et al (2020) To infuse or ingest in human laboratory alcohol research. Alcohol Clin Exp Res 44(4):764–776
40. Ramchandani VA et al (1999) A preliminary study of acute responses to clamped alcohol concentration and family history of alcoholism. Alcohol Clin Exp Res 23(8):1320–1330
41. Morzorati SL et al (2002) Self-reported subjective perception of intoxication reflects family history of alcoholism when breath alcohol levels are constant. Alcohol Clin Exp Res 26(8):1299–1306
42. Blekher T et al (2002) Saccadic eye movements are associated with a family history of alcoholism at baseline and after exposure to alcohol. Alcohol Clin Exp Res 26(10):1568–1573
43. Neumark YD et al (2004) Alcohol dehydrogenase polymorphisms influence alcohol-elimination rates in a male Jewish population. Alcohol Clin Exp Res 28(1):10–14
44. Roh S et al (2011) Role of GABRA2 in moderating subjective responses to alcohol. Alcohol Clin Exp Res 35(3):400–407
45. Marshall VJ et al (2014) Evaluation of the influence of alcohol dehydrogenase polymorphisms on alcohol elimination rates in African Americans. Alcohol Clin Exp Res 38(1):51–59
46. Yang BZ et al (2017) GRIK1 and GABRA2 variants have distinct effects on the dose-related subjective response to intravenous alcohol in healthy social drinkers. Alcohol Clin Exp Res 41(12):2025–2032
47. Ramchandani VA et al (2002) Recent drinking history: association with family history of alcoholism and the acute response to alcohol during a 60 mg% clamp. J Stud Alcohol 63(6):734–744
48. Sloan ME et al (2022) The effects of acute alcohol administration on circulating endocannabinoid levels in humans. Addict Biol 27(5):e13197
49. Ueno F et al (2022) Influence of alcohol and acetaldehyde on cognitive function: findings from an alcohol clamp study in healthy young adults. Addiction 117(4):934–945
50. Fede SJ et al (2020) Alcohol effects on globus pallidus connectivity: role of impulsivity and binge drinking. PLoS One 15(3):e0224906
51. Hendershot CS et al (2015) Application of an alcohol clamp paradigm to examine inhibitory control, subjective responses, and acute tolerance in late adolescence. Exp Clin Psychopharmacol 23(3):147–158
52. Gan G et al (2014) Alcohol-induced impairment of inhibitory control is linked to attenuated brain responses in right fronto-temporal cortex. Biol Psychiatry 76(9):698–707
53. Marcus GM et al (2021) A randomized, double-blind, placebo-controlled trial of intravenous alcohol to assess changes in atrial electrophysiology. JACC Clin Electrophysiol 7(5):662–670
54. Ramchandani VA, O'Connor S (2006) Studying alcohol elimination using the alcohol clamp method. Alcohol Res Health 29(4):286–290
55. Seyedsadjadi N et al (2022) Site of alcohol first-pass metabolism among women. JAMA Netw Open 5(3):e223711
56. Vatsalya V et al (2023) Influence of age and sex on alcohol pharmacokinetics and subjective pharmacodynamic responses following intravenous alcohol exposure in humans. Alcohol 107:144–152
57. Morzorati SL et al (2002) Maintaining steady state arterial alcohol levels in rats by using a physiologically based pharmacokinetic model. Alcohol 28(3):189–195
58. Morzorati SL, Stewart RB (2005) Development of acute tolerance during steady-state arterial alcohol concentrations: a study of auditory event-related potentials in rats. Alcohol Clin Exp Res 29(3):347–352

59. Plawecki MH et al (2012) Alcohol exposure rate control through physiologically based pharmacokinetic modeling. Alcohol Clin Exp Res 36(6):1042–1049
60. Wetherill L et al (2012) Subjective perceptions associated with the ascending and descending slopes of breath alcohol exposure vary with recent drinking history. Alcohol Clin Exp Res 36(6):1050–1057
61. Czachowski CL, Samson HH (1999) Breakpoint determination and ethanol self-administration using an across-session progressive ratio procedure in the rat. Alcohol Clin Exp Res 23(10):1580–1586
62. Czachowski CL, Samson HH (2002) Ethanol- and sucrose-reinforced appetitive and consummatory responding in HAD1, HAD2, and P rats. Alcohol Clin Exp Res 26(11):1653–1661
63. Czachowski CL et al (2002) Separate measures of ethanol seeking and drinking in the rat: effects of remoxipride. Alcohol 28(1):39–46
64. Beckwith SW, Czachowski CL (2014) Increased delay discounting tracks with a high ethanol-seeking phenotype and subsequent ethanol seeking but not consumption. Alcohol Clin Exp Res 38(10): 2607–2614
65. Kaminski BJ, Weerts EM (2014) The effects of varenicline on alcohol seeking and self-administration in baboons. Alcohol Clin Exp Res 38(2):376–383
66. Samson HH, Czachowski CL (2003) Behavioral measures of alcohol self-administration and intake control: rodent models. Int Rev Neurobiol 54:107–143
67. Sinclair JD, Senter RJ (1968) Development of an alcohol deprivation effect in rats. Q J Stud Alcohol 29(4):863–867
68. Heyser CJ, Moc K, Koob GF (2003) Effects of naltrexone alone and in combination with acamprosate on the alcohol deprivation effect in rats. Neuropsychopharmacology 28(8): 1463–1471
69. McKinzie DL et al (1998) The alcohol deprivation effect in the alcohol-preferring P rat under free-drinking and operant access conditions. Alcohol Clin Exp Res 22(5): 1170–1176
70. Czachowski CL, Legg BH, Samson HH (2001) Effects of acamprosate on ethanol-seeking and self-administration in the rat. Alcohol Clin Exp Res 25(3):344–350
71. Bertholomey ML, Verplaetse TL, Czachowski CL (2013) Alterations in ethanol seeking and self-administration following yohimbine in selectively bred alcohol-preferring (P) and high alcohol drinking (HAD-2) rats. Behav Brain Res 238:252–258
72. Windisch KA, Czachowski CL (2018) Effects of group II metabotropic glutamate receptor modulation on ethanol- and sucrose-seeking and consumption in the rat. Alcohol 66:77–85
73. DeWit H, Pierri J, Johanson CE (1989) Assessing individual differences in ethanol preference using a cumulative dosing procedure. Psychopharmacology 98(1):113–119
74. de Wit H et al (1987) Individual differences in behavioral and subjective responses to alcohol. Alcohol Clin Exp Res 11(1):52–59
75. de Wit H, Doty P (1994) Preference for ethanol and diazepam in light and moderate social drinkers: a within-subjects study. Psychopharmacology 115(4):529–538
76. Davidson D et al (1999) Effects of naltrexone on alcohol self-administration in heavy drinkers. Alcohol Clin Exp Res 23(2):195–203
77. Anton RF et al (2004) Naltrexone effects on alcohol consumption in a clinical laboratory paradigm: temporal effects of drinking. Psychopharmacology (Berl) 173(1–2):32–40
78. de Wit H et al (2003) Effects of acute social stress on alcohol consumption in healthy subjects. Alcohol Clin Exp Res 27(8):1270–1277
79. de Wit H, McCracken SG (1990) Ethanol self-administration in males with and without an alcoholic first-degree relative. Alcohol Clin Exp Res 14(1):63–70
80. Krishnan-Sarin S et al (2007) Family history of alcoholism influences naltrexone-induced reduction in alcohol drinking. Biol Psychiatry 62(6):694–697
81. Cyders MA et al (2016) Gender-specific effects of mood on alcohol-seeking behaviors: preliminary findings using intravenous alcohol self-administration. Alcohol Clin Exp Res 40(2):393–400
82. VanderVeen JD et al (2016) Negative urgency, mood induction, and alcohol seeking behaviors. Drug Alcohol Depend 165:151–158
83. Stangl BL et al (2017) Exposure-response relationships during free-access intravenous alcohol self-administration in nondependent drinkers: influence of alcohol expectancies and impulsivity. Int J Neuropsychopharmacol 20(1):31–39
84. Stangl BL et al (2019) Suggestibility is associated with alcohol self-administration, subjective alcohol effects, and self-reported drinking behavior. J Psychopharmacol 33(7): 769–778

85. Vaughan CL et al (2019) The relationship between impaired control, impulsivity, and alcohol self-administration in nondependent drinkers. Exp Clin Psychopharmacol 27(3): 236–246
86. Obst E et al (2021) Sensation seeking, impulsivity, and aggression moderate sex effects on adolescent laboratory binging. Psychol Addict Behav 35(2):208–214
87. Cyders MA et al (2021) Translating preclinical models of alcohol seeking and consumption into the human laboratory using intravenous alcohol self-administration paradigms. Addict Biol 26(6):e13016
88. Stangl BL et al (2022) Pharmacodynamic determinants of hangover: an intravenous alcohol self-administration study in non-dependent drinkers. Addict Behav 135: 107428
89. Sloan ME et al (2020) High-risk social drinkers and heavy drinkers display similar rates of alcohol consumption. Addict Biol 25(2): e12734
90. Gowin JL et al (2017) Vulnerability for alcohol use disorder and rate of alcohol consumption. Am J Psychiatry 174(11):1094–1101
91. Sloan ME et al (2018) The OPRM1 A118G polymorphism: converging evidence against associations with alcohol sensitivity and consumption. Neuropsychopharmacology 43(7): 1530–1538
92. Plawecki MH et al (2022) Binge and high-intensity drinking-associations with intravenous alcohol self-administration and underlying risk factors. Addict Biol 27(6):e13228
93. Spiga R et al (1997) Human ethanol self-administration. I: the interaction between response requirement and ethanol dose. Behav Pharmacol 8(1):91–100
94. Perkins KA, Fonte C, Grobe JE (2000) Sex differences in the acute effects of cigarette smoking on the reinforcing value of alcohol. Behav Pharmacol 11(1):63–70
95. Barrett SP et al (2006) Nicotine increases alcohol self-administration in non-dependent male smokers. Drug Alcohol Depend 81(2): 197–204
96. Barrett SP et al (2008) The role of dopamine in alcohol self-administration in humans: individual differences. Eur Neuropsychopharmacol 18(6):439–447
97. Setiawan E et al (2011) The effect of naltrexone on alcohol's stimulant properties and self-administration behavior in social drinkers: influence of gender and genotype. Alcohol Clin Exp Res 35(6):1134–1141
98. Farokhnia M et al (2018) Exogenous ghrelin administration increases alcohol self-administration and modulates brain functional activity in heavy-drinking alcohol-dependent individuals. Mol Psychiatry 23(10):2029–2038
99. Stangl BL et al (2022) The motivation for alcohol reward: predictors of progressive-ratio intravenous alcohol self-Administration in humans. J Vis Exp 182. https://doi.org/10.3791/63576
100. Plawecki MH et al (2018) Sex differences in motivation to self-administer alcohol after 2 weeks of abstinence in young-adult heavy drinkers. Alcohol Clin Exp Res 42(10): 1897–1908
101. Bujarski S et al (2018) Differences in the subjective and motivational properties of alcohol across alcohol use severity: application of a novel translational human laboratory paradigm. Neuropsychopharmacology 43(9): 1891–1899
102. National Advisory Council on Alcohol Abuse and Alcoholism (2020) Administering alcohol in human studies. 1/16/2023. Available from: https://www.hhs.gov/guidance/document/administering-alcohol-human-studies

Chapter 9

Details in Conducting Human Laboratory Alcohol Administration and Self-Administration Experiments to Assess Subjective Response and Craving for Alcohol

Benjamin L. Berey, Bailee R. Wilson, Jessica Weafer, and Robert F. Leeman

Abstract

Human laboratory methods are valuable approaches to studying the etiology of alcohol misuse and interventions to ameliorate misuse because they offer advantages such as enhanced experimental control and relative efficiency compared to other methods. Yet, published reports from laboratory alcohol administration experiments typically do not provide a sufficient degree of detail for readers to replicate these procedures themselves. The primary goal of this chapter is to provide fine-grained details on experimental laboratory methods to assess subjective response and craving for alcohol: two mechanisms underlying alcohol misuse that we have learned the most about using laboratory alcohol administration paradigms. The detailed steps described here relied heavily on one research group's approach with several alternatives offered. We describe the materials needed to conduct laboratory alcohol administration research followed by a step-by-step description of the methods involved in chronological order, beginning with day before procedures to prepare for a laboratory session; day of procedures, including calculating fixed doses of alcohol to target particular breath alcohol levels, pre-session baseline, beverage administration procedures; and day-after procedures. We conclude this chapter by discussing several key decisions investigators must make when designing a laboratory alcohol administration study. While there are several validated and reliable methods for conducting experimental laboratory research, investigators interested in subjective response and alcohol craving must carefully consider how their research questions and approaches align, while also taking into account the many potential logistical limitations.

Key words Experimental procedures, Simulated bar laboratory, Alcohol dose, Placebo beverage, Stimulation, Sedation, Urge

1 Introduction

Human laboratory methods are valuable approaches for etiology and intervention research due to enhanced experimental control and relative efficiency compared to other methods. Human laboratory studies may involve alcohol administration (also known as "alcohol challenge"; i.e., administration of a fixed alcohol dose,

Melissa A. Cyders (ed.), *Translational Research Methods for Alcohol Use Disorders*, Neuromethods, vol. 201,
https://doi.org/10.1007/978-1-0716-3267-3_9,

typically tailored to factors including participant sex and weight, in order to target a particular breath/blood alcohol concentration) or alcohol self-administration (i.e., participants have the ability to decide their own magnitude of alcohol consumption with an upper limit for safety) [1]. Alcohol self-administration may be used on its own or in combination with a fixed dose. When a fixed dose is combined with self-administration, it is often referred to as an "alcohol prime" or "priming dose," which can vary in magnitude from a low dose to ensure that all participants have at least minimal exposure (e.g., targeting a blood/breath alcohol concentration of 0.02%) or a more substantial dose (e.g., targeting a blood/breath alcohol concentration of 0.05–0.06%) [2]. Pragmatically, human laboratory alcohol administration/self-administration studies typically require less time and fewer resources than large observational studies or randomized control trials, thus these studies hold the promise of at least initial answers to research questions in a relatively efficient manner. Enhanced experimental control possible with laboratory approaches makes them particularly well suited to learning about mechanisms underlying relationships between predictors (e.g., a family history of alcohol problems [3]) and substance-related outcomes [4, 5]. Given their advantages, laboratory approaches are also well suited to determining whether new interventions have initial efficacy and sufficient promise to advance to randomized controlled trials [6], which are typically costly and time-consuming.

Subjective response and craving are two of the mechanisms underlying alcohol misuse that we have learned the most about using laboratory alcohol administration paradigms [3, 7–15]. In this chapter, we will describe in detail materials, methods, and procedures involved in conducting laboratory alcohol administration/self-administration methods to learn more about subjective response and craving. Research on subjective response and craving lends itself particularly well to alcohol administration research, because investigators' goals are often to determine level of subjective response and craving at a particular dose as opposed to self-administration where participants will receive varying doses. Thus, while we will frequently reference both, we will place greater emphasis on administration methods. Oral alcohol is the most common approach in alcohol administration/self-administration studies though intravenous administration is also an option, which offers several advantages over oral (*see* [1]). Given that oral is more common, we opted to base the methods described in this chapter on an example study involving oral administration and will focus primarily on this approach.

Published reports of these types of studies in scholarly journals typically include a degree of detail that provides a reader with a sense of the steps taken by investigators, but not enough detail to actually conduct one of these studies. In the current chapter, unlike a journal publication, we will provide a degree of detail that would enable investigators to conduct this type of study. This detailed presentation will focus primarily on how one of our research groups typically conducts this research (Leeman and colleagues with aspects of this approach adapted from Will Corbin and Stephanie O'Malley's research groups); however, there are many decision points where investigators may choose a different option. Aspects of the approach used by Weafer and colleagues will also be presented as alternatives to some of the choices made by Leeman and colleagues.

Key decisions include the type of alcohol one opts to administer, the magnitude of dose, the context in which to conduct laboratory drinking sessions, and the self-report assessments one chooses. We have attempted to identify those decision points whenever possible. Following this detailed presentation, we will discuss the pros and cons of many of these decisions along with ways in which one's research question may inform various methodologic decisions. We have opted to focus on subjective response and craving, given their importance to alcohol research and how commonly these constructs are assessed in laboratory alcohol administration/self-administration studies. Other variables pertinent to alcohol use and misuse are frequently assessed in the laboratory context as well, including mood and alcohol's effects on aspects of cognition (e.g., memory, inhibitory control).

The approach we will detail involves use of a simulated laboratory designed and decorated to resemble a bar (i.e., a bar lab; e.g., [16–18]). Other options for context include an actual bar (e.g., [19–21]) a room laid out to be a comfortable space for participants to consume alcohol, typically within an investigator's research/laboratory space (e.g., [13, 22]) and a hospital research unit (e.g., [23, 24]). Choice of context may depend on one's research question and/or pragmatics (i.e., the type of space that is available to the researcher). Though some of the methodological details described pertain only to studies occurring in a bar lab, most of these details will apply to studies occurring in other contexts.

2 Materials

Study staff use a 108-liter dry gas tank from Intoximeter to calibrate multiple handheld Alco-Sensor® VXL breathalyzers that will be used in the session. At least two stopwatches are synchronized for use by study staff. Given the bar lab context, we use music as part of our approach to make the context resemble an actual drinking situation to the extent possible. To that end, we connect an iPod or iPhone to speakers in the simulated bar laboratory. We use urine drug and pregnancy tests (if female) to confirm the absence of recent substance use (except for THC) and that female participants are not currently pregnant. In a research room adjacent to the simulated bar laboratory, we use a vitals machine to measure participants' heart rate and blood pressure. Afterward, we provide participants with a 12-ounce glass of water and a light snack (i.e., low-fat meal with 15% of calories from fat based on participant weight) [9]. Different research groups have different approaches to this snack. Ours and the Corbin group use sourdough pretzels, while the Weafer group provides granola bars. Andrea King's group serves a small meal made up of up to five components (i.e., oranges, chocolate, cheese, snack crackers, and soda) with substitutions possible according to participant food allergies and taste preferences. In all of these scenarios, the amount of food provided is tailored to participant weight. Participants also complete baseline assessments of subjective response, craving, and other constructs.

The bar lab context includes various alcohol-related cues such as beer, wine, martini, and other cocktail glasses and tap handles. In addition to music, we play television shows on a flat screen TV with a DVD player to further match the context to an actual drinking situation. The simulated lab is also equipped with a sink, refrigerator, cabinets, and ice machine behind the bartop. For our studies involving fixed-dose alcohol administration, alcoholic drinks are vodka-based using typical 80-proof vodka and a mixer (i.e., lemon-lime soda, cranberry juice, lime juice; [25]). In studies that include use of a placebo beverage that includes only trace amounts of alcohol, before utilizing glasses, we place them on trays covered with vodka-soaked towels and spray the bar down with vodka to provide olfactory cues for alcohol (described further below in Subheading 2).

3 Methods

3.1 Day-Before Procedures to Prepare for Laboratory Alcohol Administration Sessions

It is important to first verify that there are adequate supplies in the days leading up to the experimental session. These materials include a sufficient amount of vodka; flat tonic water for placebo beverage; cranberry juice, lemon–lime soda or other mixer; two lime juice bottles; and food for participants during the session. Our group empties the contents from one of the lime juice bottles and fills it with vodka to be used as a float during placebo beverage administration sessions, while the other bottle containing actual lime juice is used during alcohol administration sessions. Afterward, study staff check for sufficient hard copies when paper forms are used and make plans with the participant for transportation (e.g., Uber) to the laboratory on the day of the session. Local rules and institutional review boards differ somewhat regarding requirements to provide transportation for research participants. Our group has always provided transportation to and from the session; however, other groups have been permitted to take a more relaxed approach, including allowing participants to secure their own transportation (e.g., from a friend or family member). After confirming transportation plans, study staff send a confirmation e-mail to the participant confirming all details about the session and providing necessary reminders. One of the most important types of reminders often involves food consumption. With the exception of the aforementioned small snack or meal at the beginning of a session, most investigators request that participants abstain from food at least 4 h before alcohol administration/self-administration begins. More food in one's system to line the stomach slows down stomach emptying, leading alcohol to be less readily absorbed into the bloodstream [26, 27]. As a result, a given alcohol dose, calculated to target a particular blood alcohol concentration (BAC), will likely lead to a lower BAC than intended on a full stomach, thus affording investigators less opportunity to observe target phenomena in the lab. On the other hand, provision of a light meal/snack shortly before alcohol consumption, will increase the likelihood that alcohol administration/self-administration will be comfortable for the participant, reducing the likelihood of stomach discomfort resulting from consumption of a substantial amount of alcohol without any food in one's system. Participants are also reminded to abstain from alcohol and other recreational drug use for at least 24 h prior to the session. A simple urine drug test can confirm participant abstinence from most recreational drug use. A baseline breath alcohol test can confirm abstinence from alcohol though not for a full 24 h, thus short-term alcohol abstinence is partially verifiable and partially on the "honor system."

3.2 Day of Procedures for Conducting Laboratory Alcohol Administration Sessions

Prior to the participant arriving at the laboratory for their beverage administration session, study staff review roles and logistics for the session and verify that (a) the vitals machine is set up and working; (b) the iPod or other music device is fully charged and connected to the speakers in the simulated laboratory; (c) two stopwatches are synchronized with spare batteries; (d) all glassware in the simulated bar laboratory is clean; (e) a plan is in place to schedule the participant's transportation to the research facility; (f) the light snack with 15% of calories from fat based on participant weight has been prepared; and (g) that the correct vodka and lime juice bottles are placed in the refrigerator in the simulated laboratory. Next, study staff ensure that the breathalyzers are working properly, calibrated correctly, and there are at least 30 breathalyzer tubes ready for the session. Study staff then update a log documenting the time that participant transportation was scheduled and notify the participant via text message or phone call that their ride is on the way. Study staff also prepare a drink chart to guide drink mixing; drink log for the session to record precisely when each component of the alcohol dose was administered, when the participant took their first and last sips of each alcohol dose component; and two copies of the session script. This assumes use of paper forms (one copy for the supervisor; one copy for the research assistant and study bartender). The session script contains step-by-step details for the entire session and tasks to be completed by the session supervisor; "bartender," who serves the alcohol or placebo dose to the study participant; and research assistant(s). It is critically important that detailed, written instructions are provided to all study staff and that all study staff are trained thoroughly in their tasks prior to assisting with sessions to make sure all aspects of the study protocol are carried out accurately and consistently. These documents are labeled according to participant ID number, date, and session number.

3.2.1 Methods to Calculate Fixed-Dose Alcoholic Beverage Content

We use formulae initially derived by Watson [28], updated by Curtin & Fairchild [29] and Jatlow and colleagues [30] to calculate the overall volume of alcohol and mixer (mL) for each individual participant. This formula calculates the alcohol dose to target a specific breath alcohol concentration (BrAC) for each participant based on their (1) height; (2) weight; (3) sex; (4) total body water [TBW]; (5) duration of beverage consumption (DBC); (6) time to peak BrAC (TPB); and (7) rate of alcohol metabolism (RAM). We express time variables (#5 and #6) in terms of hours even though alcohol administration typically takes less than 1 h. *Equation 1* below demonstrates how to calculate the overall alcohol volume in grams (g) for 190-proof ethanol (i.e., medical-grade alcohol). Note that in our research and others (e.g., [29, 30]) participants' average alcohol metabolism rate is set to 0.015 g/100 mL/h. Gender-specific TBW values are derived from Watson's [28] formula (*see Eqs. 2 and 3*). For example, in a study where participants have 30 min to consume the alcoholic beverage(s) targeting a peak BrAC = 0.08% (i.e., legal intoxication in the U.S.A.) 1 h after the end of beverage administration, a 30-year old male who weighs 175 pounds and is 68 inches tall would receive 57.5 g of 190-proof ethanol (*see Eq. 4*). To convert the alcohol volume from g to mL, multiply the dose (g) in Eq. 4 by 1.27 (*see Eq. 5*). To convert the alcohol volume (mL) from 190-proof ethanol to 80-proof ethanol (e.g., Absolut® vodka), divide the value in Eq. 5 by 0.40 (*see Eq. 6*). Researchers often use different non-alcoholic mixers (e.g., juice, carbonated soda, sweetened beverages) and in various alcohol-to-mixer ratios, depending on the type of alcohol administered (e.g., [10, 22, 25, 31]). However, our group uses a 1:2 ratio of 80-proof Absolut® vodka to non-alcoholic mixer (*see Eq. 7*). Similar to prior research conducted by Cronce & Corbin [25], our group uses a non-alcoholic mixer comprising 60% lemon–lime soda, 35% cranberry juice, and 5% lime juice from concentrate (*see Eq. 8*).

$$\text{Alcohol dose (g)} = (10^{*}\ \text{BrAC}^{*}\ \text{TBW})/0.8 + 10^{*}\ \text{RAM}^{*}\ (\text{DBC} + \text{TPB})^{*}\ (\text{TBW}/0.8) \quad (1)$$

$$\text{Men's TBW} = 2.447 - 0.09516^{*}\ \text{age} + 0.1074^{*}\ \text{height (cm)} + 0.03362^{*}\ \text{weight (kg)} \tag{2}$$

$$\text{Women's TBW} = -2.097 + 0.1069^{*}\ \text{height (cm)} + 0.2466^{*}\ \text{weight (kg)} \tag{3}$$

$$\text{Alcohol dose (57.5 g)} = (10^{*}\ 0.08^{*}\ 44.9)/0.8 + 10^{*}\ 0.015^{*}\ (0.5 + 1)^{*}\ (44.9/0.8) \tag{4}$$

$$\text{Alcohol dose (73.0 mL)} = 57.5\ \text{g}^{*}\ 1.27 \tag{5}$$

$$80 - \text{proof alcohol dose (182 mL)} = 73.0\ \text{mL (190– proof pure grain alcohol)}/0.40 \tag{6}$$

$$\text{Mixer dose (364 mL)} = 182\ \text{mL}\ (80 - \text{proof alcohol dose})^{*}\ 2 \tag{7}$$

$$\begin{aligned} \text{Mixer} \quad & (218\ \text{mL of lemon} - \text{lime soda}) = 364\ \text{mL}^{*}0.60 \\ & (127\ \text{mL cranberry juice}) = 364\ \text{mL}^{*}0.35 \\ & (36\ \text{mL lime juice}) = 364\ \text{mL}^{*}0.05 \end{aligned} \tag{8}$$

A key advantage to using the validated alcohol dosing procedures described above [29, 30] is that they account for several relevant individual differences (e.g., age, height, weight, gender-specific TBW) known to affect alcohol pharmacokinetics (i.e., absorption, distribution, and elimination). However, other factors contributing to individual differences in alcohol absorption and metabolism, such as prior alcohol exposure, are not accounted for by these dosing procedures. This is a noteworthy limitation and ongoing challenge in this line of research because some individuals (often those who drink alcohol more heavily) absorb and metabolize alcohol faster than others [32–34]. Typically, a research group using these types of methods carefully will closely approximate their targeted, peak BrAC in a full sample, even a sample of relatively small size. However, there will tend to be substantial variability around that mean due mainly to individual differences in alcohol metabolism (*see* Fig. 1 in [1]).

3.2.2 Pre-Session Baseline

Once participants arrive at the research facility, we first take a breathalyzer reading using the Alco-Sensor® VXL to verify participants' BrAC (must = 0.00%). Following the breathalyzer reading, all participants provide a urine sample for drug and pregnancy testing (if female). Next, study staff provide participants with the aforementioned light meal/snack and water. After finishing the snack and water, study staff conduct semi-structured interviews

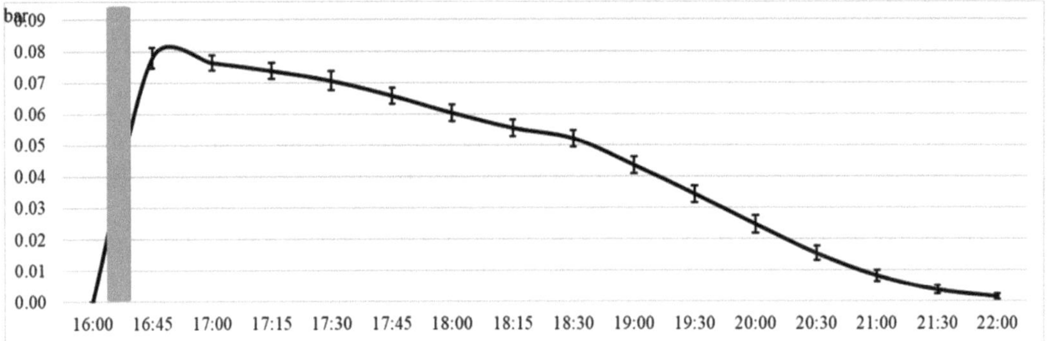

Fig. 1 Participants' mean BrACs across time during alcohol drinking session. BrAC along *Y*-axis; time point along *X*-axis. Gray bar denotes 30-min beverage administration period. Bars at each time point represent standard mean error

with participants concerning their menstrual cycle (if female). Primarily for safety reasons, we also assess concurrent medications, possible current suicidal ideation (Columbia Suicide Severity Rating Scale [CSSRS; [35]), and any adverse physical and/or mental symptoms they may have experienced in the prior week (Systematic Assessment for Treatment Emergent Effects [SAFTEE; [36]).

Study staff then review general session instructions with the participant. The general instructions cover the following information: (1) the duration of the drinking portion of the session; (2) the types of (non)alcoholic beverages that participants will have access to prior to, during, and after the drinking portion of the session; (3) how much time participants will have to consume each beverage, including a statement asking participants to "*consume each beverage as you normally would, and try your best to space your drink out over each 10-minute period*" (or whatever period the investigator has specified); (4) whether personal electronics can be used before, during, and/or after the drinking portion of the session; (5) rules regarding nicotine use before, during, and after the drinking portion of the session; (6) arrangements for when participants will receive compensation for study participation; and (7) procedures for dismissal at the end of the session. Study staff read the statement to participants described above about the pace of beverage consumption (*see* general instruction #3) in an attempt to reduce individual variability in drinking topography and alcohol pharmacokinetics. Most of these points were raised with the participant previously during the informed consent process during screening at the beginning of their study participation. However, in many cases, multiple days or longer may have elapsed between initial screening and actual session participation, and thus participants easily may forget. It is critically important both ethically and for the validity of the science that participants understand fully the rules for the session. We typically then schedule a time for a next-day phone call to confirm that participants have not experienced any adverse events following the experimental session.

Next, participants complete baseline self-reports of subjective response (e.g., BAES [37]; modified SEAS [38]) and craving in a research room immediately prior to the start of the beverage administration portion of the session. It is valuable to obtain baseline data before alcohol administration/self-administration on the day of the session to account for any effects of the participant's day up to that point (i.e., they may have had a particularly happy or challenging/stressful day which may have an impact on self-reports of subjective response and/or craving). However, not all self-report measures of subjective response are well suited to baseline testing because they explicitly link the effects described to alcohol (e.g., "The following adjectives describe feelings that some people have after drinking alcohol" [37]), which participants have not consumed yet at baseline (*see* section below on subjective response measures). After completing baseline self-reports, study staff secure the participant's personal electronics in a locked file cabinet in the research room for the duration of the drinking portion of the session. Based on preference and/or the nature of the research question, investigators may prefer to limit participant access to their electronics and other personal effects beyond just the alcohol administration/self-administration portion of the session.

While a research assistant is conducting the pre-session baseline with the participant in the research room, another study staff person (often the individual in a supervisory role for the session) combines the mixer ingredients in the simulated bar laboratory. Afterward, the supervisor places the prepared mixer back into the refrigerator behind the bar and removes the vodka and lime juice bottles, depending on beverage condition (i.e., alcohol or placebo). These steps are taken to ensure double blinding (i.e., that neither the research participant nor the study staff working directly with the participant are aware of the study condition assignment). Immediately before the start of the drinking portion of the session, the study supervisor checks that: (1) the thermostat in the simulated laboratory is set to a comfortable temperature; (2) all overhead lights are turned off and all alternative lighting (e.g., lamp, neon sign, and artificial candles on the bartop) is turned on to enhance the bar-like atmosphere; (3) copies of the session script, drink log, and drink chart to guide pouring procedures are placed on the countertop behind the bar for the study bartender; (4) three glasses are placed on the back bar; and (5) synchronized stopwatches are placed behind the bar for the study bartender and in the adjacent research room for the study supervisor. Next, the study supervisor rims the three glasses on a vodka-soaked tray, mists the inside of each glass with one spray from a spray bottle filled with vodka, mists the entire simulated bar laboratory with eight sprays of vodka, and wipes down the bartop with a vodka-soaked rag.

These procedures are implemented to increase the credibility of placebo beverage administration by providing the participant with relevant olfactory cues. Next, the study supervisor turns on the music in the simulated bar laboratory via an iPod connected to Bluetooth-enabled speakers and plays a preselected television show on the flatscreen television. If one elects to include episodes of television shows, it is important to vet them to avoid offensive content. We have preferred to vet and use the same episodes in all sessions throughout a given study. Similarly, for music, we will create a playlist of popular music without offensive content and have used the same list for an entire study. Once these procedures have been completed, the study supervisor opens the door signifying that the research assistant(s), study bartender, and participant can enter the simulated bar laboratory.

Key procedures occurring just before, during, and after beverage administration are summarized in Table 1.

Table 1
Summary of alcohol drinking session(s) from an unpublished alcohol administration study in which both subjective response and craving for alcohol were assessed

Time	Procedures
2:45–3:00	Participants arrive at the research facility
3:00 pm	Participants provide a breath alcohol reading, urine sample for analysis of illicit drugs, and pregnancy, if female. Vital signs are measured. Instructions for the session are reviewed with participants. Participants then complete baseline self-reports of subjective response, craving, and others. Lastly, participants complete baseline administration of a computer-based reaction time task. Concurrently, the session supervisor prepares the alcohol or placebo beverage and prepares the simulated bar lab for the administration period (e.g., sprays bar with vodka, lowers lights, puts music and TV on). Supervisor opens bar lab door when ready.
4:00 pm	Beverage administration period begins. Participants will have 30 min to consume three alcoholic beverages with 10 min to consume each drink.
4:30 pm	Beverage administration period ends. Participants have a small glass of water to clear their mouth of alcohol and wait 10–15 min before taking another breath alcohol reading.
4:45–6:30 pm	Participants will complete the same computer-based reaction time task and self-report questionnaires of subjective response, craving, and others. Participants also provide breath alcohol readings and have vital-signs taken. Assessments will take place every 15 min; however, participants do not complete all questionnaires and tasks at every assessment.
7:00 pm + hourly	Self-report questionnaires of subjective response, craving, and others, vital signs measured, and breath alcohol reading.
10 pm	Self-report questionnaires, vital signs measured, and breath alcohol reading. Released when BrAC $\leq$0.02%, sent home via study-provided transportation.

3.2.3 Beverage Administration Procedures

Like prior alcohol administration studies (e.g., [25, 39, 40]), participants in our fixed-dose alcohol administration paradigm have 30 min to consume three alcoholic beverages. The drinking portion of the session is separated into three, 10-min blocks for the participant to complete each individual beverage. The total alcohol dose is divided into three equal portions to be administered to the participant in each of these 10-minute blocks.

Once the participant enters the simulated bar laboratory, the study bartender (blind to beverage condition) removes the vodka bottle and mixer from the refrigerator and prepares each drink on the bartop in view of the participant. Guided by the drink chart that contains pouring procedures, the study bartender pours 1/3 of the total mixer volume, followed by 1/3 of the total vodka (or placebo) volume into each of the three glasses. Next, the study bartender adds three dashes of lime juice (actual lime juice for alcohol administration; lime juice bottle filled with vodka for placebo administration) on top of the first drink only and puts the remaining two portions of the alcohol dose back into the refrigerator. Immediately before beverage administration begins, the study bartender reminds the participant to "*please try your best to space your drink out over the next 10 minutes.*" Next, the study bartender records the time when the participant receives, starts, and finishes each beverage in the drink log. In the event that the participant has not finished their drink within 8 min, the study bartender provides a verbal reminder that each beverage must be finished by the end of each 10-minute drinking block. At the end of the first and second drinking blocks, the study bartender replaces the empty glass with a newly-prepared beverage that initiates the subsequent 10-minute drinking block. In Fig. 1, we have reported mean BrAC over time, using these methods from a recent, unpublished study.

In laboratory studies conducted by our group, participants complete experimental sessions in the afternoon with beverage administration beginning and ending promptly at a predetermined time (e.g., beginning at 16:00 and ending at 16:30 in the aforementioned recent, unpublished study). In this scenario, with the exception of completing self-reports and computerized tasks in an adjacent research room, participants remained in the simulated bar laboratory until 18:30.

Before participants leave the simulated lab, we ask them to indicate how many standard alcoholic beverages (i.e., one 12-ounce beer with 5% alcohol; one 5-ounce glass of wine with 12% alcohol; one mixed drink with 1.5 ounces of 80-proof spirits) they believe they consumed during the session. We also ask participants to estimate their peak BrAC during each session. These questions provide a manipulation check for the placebo condition, in comparison with the alcohol condition.

3.3 Procedures for the Day After Laboratory Alcohol Administration

Study staff make phone contact with participants the day after each alcohol drinking session to verify that participants have experienced no adverse events related to their alcohol consumption in the study. In the event that a participant experiences any adverse event or has any medical concerns due to their participation in the study, we immediately alert the study physician on the protocol to follow up with the participant.

4 Discussion

Despite a large literature on human laboratory alcohol administration paradigms, there have been few detailed, explicit guidelines provided on how to conduct such paradigms. We sought to address this gap in the literature by providing directions on how to execute this paradigm to examine subjective response and craving for alcohol. In providing these detailed instructions, we aimed to convey the need to conduct alcohol drinking sessions in a careful, consistent, step-by-step fashion to permit a valid test of the research questions posed in the study.

However, there are many points where research teams must make decisions in designing a protocol for an alcohol administration/self-administration study. Some of these decisions will come down to the preference of the research team and are often continued from study to study (e.g., the food provided in the light meal/snack). However, other decisions will relate directly to the research questions at hand. Thus several alternate choices could be made that depart from the materials and methods presented in the prior two sections. This chapter merely serves as a guide, and those referring to it in designing their own study protocols may make different choices. In the Discussion, we highlight several decision points and describe alternatives to the choices detailed in the preceding sections. Many of these alternative choices are guided by materials and methods used by the Weafer research group. Decisions we address in this section concern alcohol type used in sessions, magnitude of alcohol dose, time of day when sessions are conducted, context for sessions, instructions provided to participants, number and timing of assessments, and craving/subjective response measurement options.

4.1 Alcohol Type

Choosing which type of alcohol participants consume is a key consideration when conducting laboratory research focused on subjective response and/or craving. For example, alcohol administration studies have used several types of alcohol, ranging from beer to different cocktails with various liquors (e.g., vodka, rum, tequila, whiskey) to shots of liquor to jello shots to 190-proof ethanol (i.e., pure grain, "medical" alcohol) [9, 11, 18, 41–43]. Our group uses

80-proof vodka in our fixed-dose alcohol administration studies for two primary reasons. First, using 80-proof vodka (versus 190-proof ethanol) increases ecological validity given that almost all young adults have at least some experience consuming vodka, whereas considerably fewer have experience consuming medical-grade ethanol. Second, it is easier to implement a credible placebo beverage with vodka because it is clear, frequently consumed with a variety of different non-alcoholic mixers (e.g., soda/tonic water, juice, etc.), has a mild odor with a less strong taste compared to other liquors (e.g., tequila), and has smaller amounts of minor chemical constituents (e.g., congeners) relative to other alcohol types (e.g., red wine, whiskey) that can acutely impact subjective alcohol effects [44, 45]. However, there are also advantages to using 190-proof ethanol. Most notably, given the higher alcohol content in 190-proof ethanol, smaller volumes of liquid are needed for participants to achieve the desired peak BrAC. Such smaller volumes of liquid are desirable, as they can be consumed in a shorter time frame (*see Duration of beverage administration* below) and are generally less filling for participants.

4.2 Magnitude of Alcohol Dose

Along with type of alcohol administered, magnitude of dose is one of the most fundamental decisions when designing an alcohol administration/self-administration study. Alcohol doses can be based on a particular BAC/BrAC target (e.g., [16, 17, 31, 40]) or an amount of alcohol can be administered based on participant weight (e.g., [10, 12, 25, 41, 42, 46]). Often, one active alcohol dose condition is included; however, in some cases, dose ranging designs are implemented in which multiple active alcohol doses are administered, typically in counterbalanced order. In the unpublished study we have reported on in this chapter, we targeted a BrAC of 0.08%, which is common in alcohol administration studies (e.g., [17, 40]) since this level has meaning as the legal BAC/BrAC limit to operate an automobile. Some investigators have opted for larger doses targeting 0.10% [31], for instance. This level of 0.10% has also been used by our research group in other studies as the peak level to which participants can self-administer [18, 20, 21]. Subjective response research from Andrea King's group exemplifies the approach where a dose is selected based on participant weight. The King group has also implemented dose-ranging designs, such as a placebo-controlled study in which they utilized lower (0.4 g/kg) and higher (0.8 g/kg) doses. These doses yielded mean peak BrACs just under 0.04% and just under 0.09%, respectively [10]. This design enables one to determine the extent to which subjective response, craving, or other constructs increase in magnitude with increasing alcohol dose, or if introduction of any substantial dose of alcohol leads to equivalent levels of the outcome of interest. Recently, Dr. King's group conducted a dose-ranging study, including a substantially higher dose of 1.2 g/kg, yielding a mean peak BrAC of about 0.13%, which was well tolerated overall by participants [46].

As mentioned earlier, fixed dose alcohol administration and ad libitum (Latin for as much or as often as wanted) self-administration are sometimes combined in study designs. In these designs, a fixed dose, referred to as a "prime" in this context [2], is first administered, typically targeting a BrAC of 0.02–0.06%, followed by an opportunity for participants to self-administer alcohol subsequently (e.g., [16, 23]). This combined design enables a "best of both worlds" approach, in which one can assess subjective response, craving, or other constructs following a fixed dose and then assess the extent to which this subjective response or craving predicts subsequent alcohol consumption. When both fixed dosing and self-administration occur during the same session, the range of potential self-administration may be limited since a substantial alcohol dose (about 0.05–0.06% or higher) is typically required to yield substantial subjective response. Peak self-administration levels typically do not go beyond 0.10–0.12%, leaving a range of self-administration between 0.04–0.07%, which may be sufficient to demonstrate an effect of subjective response on alcohol consumption. This concern is likely less relevant for craving, since there is evidence that craving may be elicited with low dose alcohol primes [2].

4.3 Duration of Beverage Administration

Drinking duration is another key consideration for alcohol administration studies measuring subjective response and craving because speed of consumption impacts rate of absorption and metabolism. This is in addition to individual differences in alcohol absorption and metabolism [1, 41, 47]. Indeed, it is important that beverage(s) are consumed over relatively short durations so that participants have similar alcohol exposures, achieve target BrACs relatively quickly and as uniformly as possible. For example, most alcohol administration studies use rapid dosing procedures ranging from 5–30 min, albeit with few exceptions (e.g., [48, 49]). Likewise, experimentally manipulating the duration of beverage administration allows researchers to measure subjective response and/or craving between participants at roughly equivalent BrACs during ascending, peak, and/or descending limbs of the blood alcohol curve.

As described above, our dosing procedure is 30 min overall with three portions administered over a 10-min period each. However, other alcohol administration studies, particularly those that use 190-proof ethanol, may implement a shorter time period (i.e., $\leq$15 min) for beverage administration (e.g., [10, 12, 13, 50, 51]). For example, Weafer and Fillmore typically split the alcohol dose into two equal beverages and instruct participants to consume both beverages within 6 min [22, 50, 52]. Participants consume the first beverage in 2 min, followed by a 2-minute break, and then the second beverage is consumed in the final 2 min. In another approach, Weafer and de Wit typically divide the alcohol dose into four equal beverages and instruct participants to consume all four beverages in 15 min [51]. While some research groups conducting subjective response and/or craving research favor higher-proof alcohol over a shorter period of time (e.g., [10]), this practice is even more common among research groups testing effects of alcohol on cognitive task performance [22, 50–53]. In these studies, it is of paramount importance that the duration of time between the end of beverage administration and the beginning of task administration be precise. The observed effects of alcohol on cognition may vary depending upon even small variations in the amount of time between end of beverage consumption and cognitive task onset.

4.4 Expectancy Manipulations (Placebo [Including Type], No-Alcohol Control Beverage, or Both)

Implementing an effective expectancy manipulation is essential for laboratory alcohol administration studies focusing on subjective response and craving because it enables researchers to parse participants' expectations about the perceived effects of alcohol relative to actual pharmacologic alcohol effects [54–56]. As such, laboratory alcohol administration studies employ a variety of experimental manipulations, including placebo beverages, no-alcohol control beverages, where the participant is informed that they are not consuming alcohol (e.g., water or soft drinks), or inclusion of both conditions in the same study [57–59]. Investigators can use between-subject designs where participants are randomized to one specific beverage condition (i.e., either alcohol, placebo, or no-alcohol control) or within-subject designs where participants complete as many sessions as there are beverage conditions in a counterbalanced order (e.g., one half of the sample receives alcohol first followed by placebo and vice versa).

We use a placebo-controlled, within-subject design when conducting research on subjective response and craving. The placebo beverage that we use is based on prior laboratory alcohol administration research [17]. As described above, this placebo beverage consists of flat tonic water, a mixer (i.e., cranberry juice, lemon–lime soda, and lime juice from concentrate), and a small amount of vodka (i.e., ~3 mL) floated on top of the beverage just before it is served to the participant. The rationale for floating a negligible amount of alcohol on top of the placebo beverage immediately before serving it to the participant is twofold. First, this procedure increases the credibility of the placebo by providing olfactory cues for alcohol. Second, floating a small amount of vodka on the top of the beverage serves as a taste cue for when the participant takes their first sip of the beverage. Moreover, the flat tonic water used to make the placebo beverage is stored in empty vodka bottles as a way to introduce salient visual cues when the drinks are prepared by the study bartender in front of the participant. Of note, the empty bottles used to store the flat tonic water are the same size and brand as the vodka used during actual alcohol administration to avoid accidentally unblinding study staff and the participant to the beverage condition.

Another key step to creating an effective expectancy manipulation involves carefully constructing language in the consent form and other relevant study documents so that participants remain blind to the different beverage conditions throughout the study. For example, in our fixed-dose laboratory alcohol administration studies, we include the following language in the informed consent form:

> The amount of alcohol you receive may not be the same at both sessions. In each drinking session, you will not be given an amount of alcohol greater than what you have consumed on your own at least once in the past month. You will be able to watch the bartender pour each drink for you.

Including this language is crucial because it acknowledges that the beverage's contents may not be identical across sessions, but does not specify *by how much* they may differ. Moreover, broadly stating that participants may receive different amounts of alcohol at each session, but intentionally omitting information regarding a specific level of peak intoxication (e.g., BrAC = 0.08%) circumvents ethical issues concerning deception because participants technically receive alcohol at both sessions, but the amount of alcohol used as a float in the placebo beverage is negligible and tends not to produce a positive BrAC reading. As an alternative to a true placebo, other investigators have opted for an "active placebo" or low dose condition where the control beverage includes an amount of alcohol substantial enough to register a positive BrAC (e.g., 0.01%) [25, 39]. Advantages of this approach are that a low-dose control condition is more likely to be perceived accurately by participants as containing alcohol as compared to a placebo beverage [39]. However, a degree of design elegance is lost with this approach in that with the typical active alcohol and placebo design, both conditions trigger alcohol expectancies, but only one condition triggers a physiologic response to alcohol. However, with a low dose, active placebo control condition, this clear distinction is not present. Pragmatically, since differences in subjective response and craving between active alcohol and control conditions may be smaller with the low dose, active placebo control than with a true placebo control, use of a low dose control may lead to a reduction in statistical power, thus requiring a larger sample size.

Another approach to expectancy manipulation is to tell participants that they could receive one of several substances, including alcohol, a stimulant, a sedative, a marijuana-like drug, or a placebo [60]. For example, Weafer and de Wit [51] used the following language in the consent form: "*You will be given 4 beverages containing one of the following:*

1. *Stimulant* (e.g., *caffeine*)
2. *Sedative/tranquilizer* (e.g., *antihistamine*)
3. *Alcohol*
4. *Placebo (inactive substance)*"

This approach limits alcohol-specific expectancies, as a significant number of participants often report receiving a substance other than alcohol. This approach is especially helpful in repeated sessions designs and when the researchers want to reduce bias or expectancies associated with alcohol consumption.

4.4.1 Best Practices for Manipulation Checks

When implementing expectancy manipulations in alcohol administration studies focused on subjective response and craving, it is important to evaluate the effectiveness of that manipulation. One of the most common manipulation checks to gauge the effectiveness of a placebo is by asking participants to indicate how many standard alcoholic beverages they consumed during the session after the end of the beverage administration procedures. However, in our fixed-dose laboratory alcohol administration studies, we also ask participants to estimate their estimated peak BrAC during each session. Our rationale for including an additional item concerning BrAC is that the effectiveness of the placebo may be inflated when researchers just ask about the number of standard alcoholic beverages because participants (1) expect to consume alcohol at each session and (2) receive three beverages of equal volume during the drinking portion of the session in our approach, which participants may equate with standard drink size even though they are unlikely to be.

For example, in a pilot study conducted by our group, young adults who engaged in frequent heavy drinking ($N = 23$) completed two drinking sessions that included alcohol or placebo administration (within-subject, in counterbalanced order) targeting a peak BrAC of 0.08% 60 min following the end of beverage administration. Participants had 30 min to consume three beverages of equal volume. On average, participants in this study estimated consuming approximately three standard alcoholic beverages during the alcohol session and approximately two and a half standard alcoholic beverages during the placebo session. Accordingly, there were no statistically significant differences between the estimated number of standard alcoholic beverages consumed across conditions (Alcohol Session: $M = 3.05$, $SD = 0.79$; Placebo Session: $M = 2.57$, $SD = 1.20$) $t(21) = 1.915$, $p = 0.069$. In contrast and as expected, participants estimated achieving considerably higher BrACs during the alcohol session ($M = 0.18\%$, $SD = 0.16$) than in the placebo session ($M = 0.08$, $SD = 0.08$), $t(21) = 2.89$. $p = 0.009$, although it is notable that participants estimated receiving an intoxicating dose of alcohol during placebo beverage administration. Further, there was a considerable range of estimated

BrACs during the placebo session (range = 0.00–0.40). These results suggest that asking about BrAC may be more sensitive to expectancy manipulations than items asking about standard drinks and can provide more nuanced information concerning the effectiveness of the placebo beverage.

4.4.2 Avoiding Order Effects

Order effects are a risk associated with any within-subject design. Order effects occur when participants respond in a noticeably different way based on whether the session is the first, second, or third, etc., in which they have participated. This effect could be independent of or interactive with effects of the study condition that are the main purpose of the study. As such, order effects detract from an investigator's ability to detect an effect of their intervention or other experimental manipulation on subjective response, craving, or another variable. When order effects occur, it can be difficult to determine why.

There are a few approaches that can be used to prevent order effects. Counterbalancing is perhaps the most basic step taken to avoid order effects. While an important step, counterbalancing will not prevent order effects on its own. In within-subjects designs, sessions typically occur at least two or more days apart in order to avoid effects of the first session impacting the second, due to lingering effects of alcohol from the first session to the second in cases of high alcohol doses, or perhaps fatigue from a long first experimental session affecting responses on the second. Lengthy sessions are often difficult to avoid as per NIAAA guidelines [61] for alcohol administration, investigators must strongly encourage participants to remain in the research facility until their BrAC reaches a safe level, typically between 0.02% and 0.04%. With high alcohol doses in particular, this process can take up to several hours. However, research groups can control most aspects of participants' experience during experimental sessions. Making sessions as pleasant as possible for participants can help to avoid negative carryover effects across sessions due to participants having an aversive experience in a session. Investigators can improve the participant experience by providing comfortable spaces to spend time (e.g., comfortable chairs), snacks and drinks, and recreation opportunities (e.g., games, television with DVDs or streaming service, reading materials) that do not detract from the investigator's ability to address their research questions. Order effects may be due to the novelty of the experience of participating in an alcohol drinking session. Decline in novelty from the first session to the next could manifest in diminished ratings of stimulant response to alcohol, for instance. This novelty effect could be diminished by exposing participants to the space where alcohol administration/self-administration is to take place before the session, perhaps during an in-person screening or orientation appointment. If participants take this step, our recommendation is to "activate" the space so that the environment mirrors how it will be during an actual alcohol drinking session (e.g., lower lighting, turn music on).

Though this will add time and expense, investigators may consider a pre-session where participants are administered a small amount of alcohol (e.g., targeting a BrAC of 0.02%), so that they become accustomed to consuming alcohol in this space before the main experimental sessions.

4.5 Time of Day

Depending on the study and research laboratory, alcohol administration studies are conducted at different times of the day [47]. For example, in some research studies, participants consume alcohol in the morning (e.g., [14, 15, 62]), whereas in others, participants consume alcohol in the afternoons or evenings (e.g., [9, 16, 18, 63]). The time when participants consume alcohol may influence their subjective response and craving [64, 65] to a greater extent than their actual drinking behaviors (e.g., ad libitum alcohol self-administration) [66]. One prior study enrolled a small sample of healthy adult males ($N = 12$) to consume a moderate dose of alcohol at two different times of day (i.e., 9:00 am and 5:00 pm). Results indicated that participants experienced greater subjective sedation after consuming alcohol in the morning versus afternoon [65]. Though subjective response to alcohol is defined as individual differences in response to alcohol's pharmacologic effects, it is difficult to disentangle entirely pharmacologic response from expectancies, even when effective placebo doses are used [47]. Given that alcohol is more commonly consumed later in the day, it seems likely that even those who drink heavily might be less likely to crave alcohol earlier in the day. Decisions regarding time of day may relate to pragmatics (i.e., when suitable staff and/or space to conduct sessions is available). If sessions require skilled staff (e.g., for blood draws), this may limit the time of day when one can conduct sessions. Investigators may emphasize or deemphasize the importance of time of day based on their research questions. In particular, if one is measuring constructs where alcohol expectancies are not likely to be relevant, time of day may not have a substantial effect on the results of the study.

4.6 Context

The physical context (room, lighting, music, etc.) and social context (fellow drinkers, bartender, alone, etc.) of drinking situations may affect participant decisions regarding number of drinks consumed and speed of consumption [67, 68]. Thus, both physical and social context are important considerations in designing alcohol administration and self-administration studies.

Alcohol administration/self-administration sessions can be conducted in a variety of environments, including clinics or hospital research units (e.g., [23]), actual bars (e.g., [19]), or simulated "bar labs" ([17]). Additionally, items introduced within the physical environment can vary and may act as cues for alcohol (i.e., red solo cup containing alcohol) and can prompt craving. Both aspects of physical context may have an impact on participants' subjective response and craving [17, 69].

Additionally, the social aspects (i.e., who one is with) of a drinking occasion can influence subjective response and craving for alcohol [70]. Alcohol administration/self-administration sessions most often involve an individual participant only (e.g., [23, 31]). For studies with research questions in which social interaction is paramount, investigators may opt for a confederate: a study staff member who purports to be a research participant, but acts in a scripted manner in alignment with the research team (e.g., [71]). Protocols including confederates can be challenging to get approved by IRBs in the United States given that they involve deception that may influence and may even be designed to influence participant alcohol self-administration in the lab. This social influence may impact participant consumption behavior after the session. To guard against this, careful debriefing after the session is essential. To add a social element to alcohol administration sessions with a primary focus on subjective response to alcohol, Andrea King's group has a study staff member sit with the participant during administration and engage in conversation. The participant is aware that this individual is a study staff member, thus they are not a confederate (e.g., [13]). The clearest way to ensure a social element is to conduct alcohol administration/self-administration sessions among small groups of participants, often 2–4 at a time (e.g., [16–18, 20, 21, 40]). This approach offers ecological validity in that individuals, especially young adults, typically consume alcohol in groups; however, this approach also introduces complexity in multiple respects. While it is beneficial to introduce a social element, it may be disadvantageous if this social element becomes so strong that it is the primary driver of participants' ad libitum self-administration decisions (i.e., how much alcohol to self-administer), unless this is the primary goal based on the research question. For this reason, in our research group, we prefer running sessions with three participants at a time rather than with pairs because, anecdotally, sessions involving two participants tend to have the highest degree of similarity in the amount of alcohol self-administered by participants. Due to the effects participants have on one another, when sessions are conducted within small groups, group membership must be accounted for analytically, often with the use of hierarchical linear modeling, in which individual participants are nested within groups (e.g., [17, 18]).

4.7 Instructions and Methods to Increase Adherence

To minimize effects of confounding variables, careful and detailed instructions are necessary [72]. Such instructions can include restricting eating up to 4 h before alcohol consumption, requesting that participants not drive themselves to drinking sessions, and restricting alcohol and drug consumption before the session.

However, only prior alcohol and drug consumption can be validated reliably with a simple breathalyzer and urine drug test, respectively. Given alcohol's short half-life of only 4–5 h, a breathalyzer or blood alcohol reading can only account for alcohol consumption over the past few to several hours, depending on the amount and type of alcohol consumed and speed of consumption. Thus, lengthier periods of alcohol abstinence will be challenging to confirm without more sophisticated testing of ethylglucuronide (EtG) [30] or phosphatidylethanol (PEth) [73]. Even then it is debatable as to whether these tests provide definitive confirmation of alcohol abstinence over a period lasting longer than several hours. Participant observation over the period when abstinence from food or alcohol is requested is the only way to know definitively that participants adhered. However, such observation is time- and resource-intensive, particularly when confirming several hours of abstinence and may have a negative effect on the number or diversity of participants one can recruit into a study. Use of a bogus pipeline (i.e., claiming that adherence can be confirmed when it cannot be) is a possibility, though requires a degree of deception and must be sufficiently convincing to participants [74]. A manipulation check will be needed afterward to determine whether participants believed the investigators' claim. Ultimately, many investigators rely upon participant adherence on an "honors system" basis, given these logistical challenges in ensuring adherence.

4.8 Number and Timing of Assessments

The number and timing of assessments evaluating subjective response and craving across the blood alcohol curve vary across alcohol administration studies [2, 47]. An absorption period, typically between 10–15 min, is common after the end of alcohol administration before the first assessment takes place (e.g., [39]). After the absorption period, investigators will administer assessments according to one of two approaches. The more straightforward approach is to space assessments out by standardized periods of time (e.g., 15–30 min) that typically increase later in a session [2]. The other main approach is to anchor assessments to particular target BrAC/BACs (e.g., [31]). This approach could be more challenging for multiple reasons. Investigators may have to take multiple readings before participants' BrACs approximate the target level. In some cases, participants may not reach the exact target BrAC. There may be an added degree of participant fatigue as well if additional BrAC readings are needed. As a result of this approach, participants will provide assessments at different times given individual differences in rates of alcohol absorption and metabolism.

In terms of the number of assessments, a balance must be struck. For example, it is important to capture changes in craving and different types of subjective responses (e.g., stimulation and sedation) across the ascending and descending limbs of the blood alcohol curve. As such, at a minimum, we recommend that researchers measure subjective response and alcohol craving at baseline and after alcohol administration while BrACs are rising, at peak, and declining. Equally important, having participants complete too many assessments may lead to fatigue, irritation, reactivity, and/or frustration that may affect the validity of the responses.

4.9 Measures of Subjective Response

There are several subjective response measures used in human alcohol laboratory administration paradigms, each with its own corresponding strengths and limitations. A commonality across most of these measures is that the individual items tend to focus exclusively on the extent to which alcohol has produced a given subjective effect (e.g., "difficulty concentrating," "excited" [37]). This dovetails with the definition of subjective response concerning individual differences in pharmacologic response to alcohol [47]. In contrast, items in alcohol expectancy measures tend to be more elaborate, making explicit that individual perception or opinion is involved rather than directly measuring acute alcohol effects (e.g., "When I drink alcohol, I expect that ... It would be easier to express my feelings" [75]). Given the focus of this chapter, we have briefly touched upon several of the most common measures. Given that our primary focus is on providing details of study procedures rather than discussion of advantages and disadvantages of subjective response measures, an extensive discussion is outside the scope of this chapter.

The Biphasic Alcohol Effects Scale (BAES) is a 14-item unipolar adjective rating scale. Items are rated on an 11-point scale anchored by "not at all" and "extremely." The BAES includes subscales capturing stimulation (e.g., elated, energized) and sedation (down, sluggish) [37]. However, the BAES does not comprehensively assess all subjective alcohol effects (e.g., negative, stimulating effects like "aggressive" or positive, sedating effects like "mellow") or account to a great extent for physiological aspects of subjective response such as physical sensation (racing heart, light-headedness, etc). While the instructional set for the original BAES asks participants to "please rate the extent to which *drinking alcohol* has produced these feelings in you at the present time," other versions have modified the instructional set to remove any mention of alcohol by asking participants to "rate the extent to which these words describe your feelings at the present time" [76]. Because the original version asks participants to rate the extent to which drinking alcohol has produced a set of feelings at present, it does not allow for any baseline, pre-drinking assessments (Martin et al., 1993). While there is intuitive appeal to ascertaining subjective mood states prior to and following drinking with the same measure, by not asking participants about different subjects effects due to drinking alcohol, it is possible that participants could attribute certain subjective responses to a combination of alcohol effects (both pharmacological and expectancy-based) and other extraneous factors (e.g., physical context). Thus, it is germane to tie together subjective responses to the effects of alcohol, especially in studies with repeated measures, as this may distinguish pharmacologic differences between alcohol and placebo sessions.

The Subjective Effects of Alcohol Scale (SEAS) is another 14-item unipolar adjective rating scale. Items are rated on an 11-point scale with "not at all" and "extremely" anchors on the poles with a midpoint anchor (i.e., "moderately"). The SEAS incorporates all four possible combinations of positive/negative emotional valence and stimulant/sedative factors, whereas the BAES does not [38]. Specifically, the SEAS captures subjective alcohol effects that are grouped together by valence (i.e., positive or negative) and arousal (i.e., stimulating or sedating) to incorporate low arousal positive (e.g., relaxed), low arousal negative (e.g., woozy), high arousal positive (e.g., talkative), and high arousal negative (e.g., rude) effects. Similar to the BAES, the SEAS states that "the following words describe feelings that are sometimes produced by drinking alcohol. Please rate the extent to which you are feeling each of the following effects at the present time." Because the SEAS does not explicitly instruct participants to indicate how drinking alcohol has produced different alcohol effects, this measure can be used prior to and following beverage administration. As a newer measure, the SEAS does not have as extensive a supporting literature as the BAES.

The Drug Effects Questionnaire (DEQ; [77]) asks participants to indicate how five different descriptors applies to them rated on 100 mm visual analog scales with anchors such as "not at all" to "extremely." The DEQ has been popularized because it quantifies a subjective response's strength (e.g., Do you feel a drug effect, right now?) and appeal (e.g., Do you [dis]like any of the effects you are feeling right now?). In contrast to the BAES and SEAS, the DEQ is not tied to a particular substance and thus offers the potential for continuity of assessment and evaluation of effects across substances that is less feasible with alcohol-specific measures. However, the DEQ has been administered in various forms across studies, for instance, with varying item anchors, thus comparison across studies is more challenging than for other measures [78].

The Addiction Research Center Inventory (ARCI) was initially developed in the 1960s at the National Institute of Mental Health to assess acute subjective drug effects [79]. The ARCI instructs participants to indicate whether each statement about a particular subjective effect is true or false (e.g., "I feel a very pleasant emptiness"). The ARCI was developed to assess subjective effects for various substances and includes the following subscales: Pentobarbital-Chlorpromazine-Alcohol (PCA), Amphetamine, Morphine-Benzedrine (MB), Lysergic Acid Diethylamide (LSD),

Benzedrine, and Marijuana. Prior studies have included the ARCI Amphetamine subscale to measure stimulating subjective alcohol responses, PCA subscale to measure sedating subjective alcohol responses, and MB to assess euphoric effects [13]. One important note is that some items appear on multiple subscales. Accordingly, items should only be included once when using multiple scales and should be scored on all relevant scales [80].

Single item measures/questionnaires are valuable in that they lessen the constraint on participant time and energy, lessening the chance for fatigue. However, in general, these measures have psychometric limitations, including moderate test-retest reliability and convergent validity [81].

Though not measures of subjective response per se, physiologic (e.g., heart rate) and biologic (e.g., cortisol) measures of arousal can be coupled with assessments of subjective response, often as means of validating subjective response measures. This methodology is beneficial in assessing underlying mechanisms of alcohol reward and tolerance [13]. An added benefit is that these objective measures can be tracked continuously (e.g., heart rate variability using a wearable fitness tracker) with relatively minimal participant burden. However, these require additional resources and participant time.

4.10 Measures of Craving

Similar to our approach to subjective response measures, while it is important to present overviews of the most commonly used craving measures along with their advantages and disadvantages, extensive discussion of these measures is outside the scope of the present chapter. There are several retrospective measures of craving that ask participants to rate their level of craving thinking back over a period of time. Given the laboratory focus of the present chapter, we have omitted these retrospective measures of craving.

Craving is often assessed using The Alcohol Urge Questionnaire (AUQ; [82]) which is a reliable and valid 8-item measure with items rated on 7-point Likert scales. Participants rate their degree of agreement with statements concerning their desire to drink, expectations of desired outcomes when drinking, and difficulty avoiding alcohol if it were available. Scores on the AUQ have been shown to be significantly related to alcohol dependence severity and self-reported quantity of alcohol use [83].

The Desires for Alcohol Questionnaire (DAQ) consists of 36 items rated on 7-point Likert scales with four subscales: intentions to consume alcohol, desires to drink alcohol, anticipation of positive outcomes as a result of drinking, and anticipation of relief from negative affect or withdrawal from drinking. Evidence supports the reliability and validity of the measure, including comparisons between individuals self-reported more and less heavy drinking, in which the latter group scored significantly higher on the DAQ [84].

In addition to full scales like the AUQ and DAQ, it is common for investigators to measure craving using just a visual analog scale item. One common approach along these lines is a single item based on the DEQ item concerning the extent that participants "want more" alcohol [2]. As an example of the validity of this approach, in our group, we found that self-reported craving assessed via a single item measure of "want more" following placebo administration in the lab significantly predicted subsequent alcohol-self administration [85].

5 Conclusion

Human laboratory methods have immense value for studying the etiology of alcohol misuse and informing interventions to ameliorate misuse. Human laboratory alcohol administration research provides enhanced experimental control and relative efficiency compared to observational and/or more ecologically valid methods (e.g., ambulatory assessments in the real world). This chapter focused on subjective response and alcohol craving because these are two key mechanisms underlying alcohol misuse that we have learned the most about using laboratory alcohol administration paradigms. Noting a limitation in the literature that journal articles typically provide limited methodological details for what are often complex procedures, we have provided a step-by-step approach to conduct human laboratory alcohol administration experiments that assess subjective response and alcohol craving. The detailed steps described here relied heavily on one research group's approach with several alternatives offered. In addition, we walked through several of the decisions investigators must make in designing one of these studies. Clearly, there is not one definitive approach for most of these steps. It is most important that investigators consider carefully each of these steps and make decisions in service of their research questions, taking into account any logistical limitations.

References

1. Cyders MA, Plawecki MH, Corbin W, King A, McCarthy DM, Ramchandani VA, Weafer J, O'Connor SJ (2020) To infuse or ingest in human laboratory alcohol research. Alcohol Clin Exp Res 44(4):764–776. https://doi.org/10.1111/acer.14305
2. Halsall L, Jones A, Roberts C, Knibb G, Rose AK (2022) The impact of alcohol priming on craving and motivation to drink: a meta analysis. Addiction 117:2986–3003. https://doi.org/10.1111/add.15962
3. Schuckit MA (1984) Subjective responses to alcohol in sons of alcoholics and control subjects. Arch Gen Psychiatry 41(9):879. https://doi.org/10.1001/archpsyc.1984.01790200061008
4. Plebani JG, Ray LA, Morean ME, Corbin WR, MacKillop J, Amlung M, King AC (2012) Human laboratory paradigms in alcohol research. Alcohol Clin Exp Res 36(6): 455–466. https://doi.org/10.1111/j.1530-0277.2011.01704.x

5. Ray LA, Green R, Roche DJ, Magill M, Bujarski S (2019) Naltrexone effects on subjective responses to alcohol in the human laboratory: a systematic review and meta-analysis. Addict Biol 24(6):1138–1152. https://doi.org/10.1111/adb.12747
6. Yardley MM, Ray LA (2017) Medications development for the treatment of alcohol use disorder: insights into the predictive value of animal and human laboratory models. Addict Biol 22(3):581–615. https://doi.org/10.1111/adb.12349
7. Bujarski S, Hutchison KE, Roche DJO, Ray LA (2015) Factor structure of subjective responses to alcohol in light and heavy drinkers. Alcohol Clin Exp Res 39(7):1193–1202. https://doi.org/10.1111/acer.12737
8. Courtney KE, Ray LA (2014) Subjective responses to alcohol in the lab predict neural responses to alcohol cues. J Stud Alcoh Drugs 75(1):124–135. https://doi.org/10.15288/jsad.2014.75.124
9. King AC, Houle T, Wit H, Holdstock L, Schuster A (2002) Biphasic alcohol response differs in heavy versus light drinkers. Alcohol Clin Exp Res 26(6):827–835. https://doi.org/10.1111/j.1530-0277.2002.tb02611.x
10. King AC, de Wit H, McNamara PJ, Cao D (2011) Rewarding, stimulant, and sedative alcohol responses and relationship to future binge drinking. Arch Gen Psychiatry 68(4):389. https://doi.org/10.1001/archgenpsychiatry.2011.26
11. King AC, McNamara PJ, Hasin DS, Cao D (2014) Alcohol challenge responses predict future alcohol use disorder symptoms: a 6-year prospective study. Biol Psychiatry 75(10):798–806. https://doi.org/10.1016/j.biopsych.2013.08.001
12. King AC, Hasin D, O'Connor SJ, McNamara PJ, Cao D (2016) A prospective 5-year re-examination of alcohol response in heavy drinkers progressing in alcohol use disorder. Biol Psychiatry 79(6):489–498. https://doi.org/10.1016/j.biopsych.2015.05.007
13. King AC, Vena A, Howe MM, Feather A, Cao D (2022) Haven't lost the positive feeling: a dose-response, oral alcohol challenge study in drinkers with alcohol use disorder. Neuropsychopharmacology 47(11):1892–1900. https://doi.org/10.1038/s41386-022-01340-2
14. Schuckit MA (1988) A simultaneous evaluation of multiple markers of ethanol/placebo challenges in sons of alcoholics and controls. Arch Gen Psychiatry 45(3):211. https://doi.org/10.1001/archpsyc.1988.01800270019002
15. Schuckit MA (2000) Response to alcohol in daughters of alcoholics: a pilot study and a comparison with sons of alcoholics. Alcohol Alcohol 35(3):242–248. https://doi.org/10.1093/alcalc/35.3.242
16. Corbin WR, Gearhardt A, Fromme K (2008) Stimulant alcohol effects prime within session drinking behavior. Psychopharmacology 197(2):327–337. https://doi.org/10.1007/s00213-007-1039-x
17. Corbin WR, Scott C, Boyd SJ, Menary KR, Enders CK (2015) Contextual influences on subjective and behavioral responses to alcohol. Exp Clin Psychopharmacol 23(1):59–70. https://doi.org/10.1037/a0038760
18. Leeman RF, Berey BL, Frohe T, Rowland BHP, Martens MP, Fucito LM, Stellefson M, Nixon SJ, O'Malley SS (2022) A combined laboratory and field test of a smartphone breath alcohol device and blood alcohol concentration estimator to facilitate moderate drinking among young adults. Psychol Addict Behav 36(6):710–723. https://doi.org/10.1037/adb0000780
19. Davidson D, Palfai T, Bird C, Swift R (1999) Effects of naltrexone on alcohol self-administration in heavy drinkers. Alcohol Clin Exp Res 23(2):195–203. https://doi.org/10.1111/j.1530-0277.1999.tb04099.x
20. Leeman RF, Corbin WR, Nogueira C, Krishnan-Sarin S, Potenza MN, O'Malley SS (2013) A human alcohol self-administration paradigm to model individual differences in impaired control over alcohol use. Exp Clin Psychopharmacol 21(4):303–314. https://doi.org/10.1037/a0033438
21. Leeman RF, Nogueira C, Wiers RW, Cousijn J, Serafini K, DeMartini KS, Bargh JA, O'Malley SS (2018) A test of multisession automatic action tendency retraining to reduce alcohol consumption among young adults in the context of a human laboratory paradigm. Alcohol Clin Exp Res 42(4):803–814. https://doi.org/10.1111/acer.13613
22. Weafer J, Fillmore MT (2013) Acute alcohol effects on attentional bias in heavy and moderate drinkers. Psychol Addict Behav 27(1):32–41. https://doi.org/10.1037/a0028991
23. Krishnan-Sarin S, O'Malley SS, Franco N, Cavallo DA, Morean M, Shi J, Pittman B, Krystal JH (2015) *N*-methyl-d-aspartate receptor antagonism has differential effects on alcohol craving and drinking in heavy drinkers. Alcohol Clin Exp Res 39(2):300–307. https://doi.org/10.1111/acer.12619
24. Kerfoot K, Pittman B, Ralevski E, Limoncelli D, Koretski J, Newcomb J, Arias AJ, Petrakis IL (2013) Effects of family history of alcohol dependence on the subjective

response to alcohol using the intravenous alcohol clamp. Alcohol Clin Exp Res 37(12): 2011–2018. https://doi.org/10.1111/acer.12199
25. Cronce JM, Corbin WR (2010) Effects of alcohol and initial gambling outcomes on within-session gambling behavior. Exp Clin Psychopharmacol 18(2):145–157. https://doi.org/10.1037/a0019114
26. Sedman AJ, Wilkinson PK, Sakmar E, Weidler DJ, Wagner JG (1976) Food effects on absorption and metabolism of alcohol. J Stud Alcohol 37(9):1197–1214. https://doi.org/10.15288/jsa.1976.37.1197
27. Sadler DW, Fox J (2011) Intra-individual and inter-individual variation in breath alcohol pharmacokinetics: the effect of food on absorption. Sci Justice 51(1):3–9. https://doi.org/10.1016/j.scijus.2010.10.003
28. Watson PE (1989) Total body water and blood alcohol levels: updating the fundamentals. In: Crow K, Batt R (eds) Human metabolism of alcohol, Pharmacokinetics, medicolegal aspects, and general interest, vol 1. CRC Press, Boca Raton, pp 41–58
29. Curtin JJ, Fairchild BA (2003) Alcohol and cognitive control: implications for regulation of behavior during response conflict. J Abnorm Psychol 112(3):424–436. https://doi.org/10.1037/0021-843x.112.3.424
30. Jatlow PI, Agro A, Wu R, Nadim H, Toll BA, Ralevski E, Nogueira C, Shi J, Dziura JD, Petrakis IL, O'Malley SS (2014) Ethyl glucuronide and ethyl sulfate assays in clinical trials, interpretation, and limitations: results of a dose ranging alcohol challenge study and 2 clinical trials. Alcohol Clin Exp Res 38(7): 2056–2065. https://doi.org/10.1111/acer.12407
31. Amlung M, McCarty KN, Morris DH, Tsai CL, McCarthy DM (2015) Increased behavioral economic demand and craving for alcohol following a laboratory alcohol challenge. Addiction 110(9):1421–1428. https://doi.org/10.1111/add.12897
32. Dubowski KM (1985) Absorption, distribution and elimination of alcohol: highway safety aspects. J Stud Alcoh Suppl s10:98–108. https://doi.org/10.15288/jsas.1985.s10.98
33. Whitfield JB, Martin NG (1994) Alcohol consumption and alcohol pharmacokinetics: interactions within the normal population. Alcohol Clin Exp Res 18(2):238–243. https://doi.org/10.1111/j.1530-0277.1994.tb00007.x
34. Jones AW (2010) Evidence-based survey of the elimination rates of ethanol from blood with applications in forensic casework. Forensic Sci Int 200(1–3):1–20. https://doi.org/10.1016/j.forsciint.2010.02.021
35. Posner K, Brown GK, Stanley B, Brent DA, Yershova KV, Oquendo MA, Currier GW, Melvin GA, Greenhill L, Shen S, Mann JJ (2011) The Columbia–suicide severity rating scale: initial validity and internal consistency findings from three multisite studies with adolescents and adults. Am J Psychiatr 168(12): 1266–1277. https://doi.org/10.1176/appi.ajp.2011.10111704
36. Johnson BA, Ait-Daoud N, Roache JD (2005) The combine SAFTEE: a structured instrument for collecting adverse events adapted for clinical studies in the alcoholism field. J Stud Alcoh Suppl s15:157–167. https://doi.org/10.15288/jsas.2005.s15.157
37. Martin CS, Earleywine M, Musty RE, Perrine MW, Swift RM (1993) Development and validation of the biphasic alcohol effects scale. Alcohol Clin Exp Res 17(1):140–146. https://doi.org/10.1111/j.1530-0277.1993.tb00739.x
38. Morean ME, Corbin WR, Treat TA (2013) The subjective effects of alcohol scale: development and psychometric evaluation of a novel assessment tool for measuring subjective response to alcohol. Psychol Assess 25(3): 780–795. https://doi.org/10.1037/a0032542
39. Corbin WR, Cronce JM (2007) Alcohol effects on behavioral control: the impact of likelihood and magnitude of negative consequences. Alcohol Clin Exp Res 31(6):955–964. https://doi.org/10.1111/j.1530-0277.2007.00389.x
40. Quinn PD, Fromme K (2016) Individual differences in subjective alcohol responses and alcohol-related disinhibition. Exp Clin Psychopharmacol 24(2):90–99. https://doi.org/10.1037/pha0000065
41. Conrod PJ, Peterson JB, Pihl RO, Mankowski S (1997) Biphasic effects of alcohol on heart rate are influenced by alcoholic family history and rate of alcohol ingestion. Alcohol Clin Exp Res 21(1):140–149. https://doi.org/10.1111/j.1530-0277.1997.tb03742.x
42. Drobes DJ, Anton RF, Thomas SE, Voronin K (2004) Effects of naltrexone and nalmefene on subjective response to alcohol among non-treatment-seeking alcoholics and social drinkers. Alcohol Clin Exp Res 28(9):1362–1370. https://doi.org/10.1097/01.alc.0000139704.88862.01
43. Ralevski E, Gueorguieva R, Limoncelli DD, Husain R, Serrita Jane J, Petrakis I (2006) Gelatin "shots" as a new method for alcohol administration in a laboratory setting. Alcohol

Clin Exp Res 30(3):473–479. https://doi.org/10.1111/j.1530-0277.2006.00064.x

44. Keane TM, Lisman SA, Kreutzer J (1980) Alcoholic beverages and their placebos: an empirical evaluation of expectancies. Addict Behav 5(4):313–328. https://doi.org/10.1016/0306-4603(80)90005-2
45. Rohsenow DJ, Marlatt G (1981) The balanced placebo design: methodological considerations. Addict Behav 6(2):107–122. https://doi.org/10.1016/0306-4603(81)90003-4
46. Vena A, Howe M, Fridberg D, Cao D, King AC (2020) The feasibility, tolerability, and safety of administering a very high alcohol dose to drinkers with alcohol use disorder. Alcohol Clin Exp Res 44(12):2588–2597
47. Morean ME, Corbin WR (2010) Subjective response to alcohol: a critical review of the literature. Alcohol Clin Exp Res 34(3): 385–395. https://doi.org/10.1111/j.1530-0277.2009.01103.x
48. Shannon EE, Staniforth ER, McNamara J, Bernosky-Smith KA, Liguori A (2011) Response inhibition impairments predict alcohol-induced sedation. Alcohol Alcohol 46(1):33–38. https://doi.org/10.1093/alcalc/agq080
49. Aston ER, Shannon EE, Liguori A (2014) Anxiety, sedation, and simulated driving in binge drinkers. Psychol Addict Behav 28(2): 359–366. https://doi.org/10.1037/a0036511
50. Weafer J, Fillmore MT (2012) Comparison of alcohol impairment of behavioral and attentional inhibition. Drug Alcohol Depend 126(1–2):176–182. https://doi.org/10.1016/j.drugalcdep.2012.05.010
51. Weafer J, Phan KL, de Wit H (2020) Poor inhibitory control is associated with greater stimulation and less sedation following alcohol. Psychopharmacology 237(3):825–832. https://doi.org/10.1007/s00213-019-05420-y
52. Weafer J, Fillmore MT (2015) Alcohol-related cues potentiate alcohol impairment of behavioral control in drinkers. Psychol Addict Behav 29(2):290–299. https://doi.org/10.1037/adb0000013
53. Gilbertson R, Ceballos NA, Prather R, Nixon SJ (2009) Effects of acute alcohol consumption in older and younger adults: perceived impairment versus psychomotor performance. J Stud Alcoh Drugs 70(2):242–252. https://doi.org/10.15288/jsad.2009.70.242
54. Bernstein MH, Wood MD, Colby SM (2016) A new paradigm for credibly administering placebo alcohol to underage drinkers. Addict Behav 52:22–27. https://doi.org/10.1016/j.addbeh.2015.08.004
55. Schlauch RC, Waesche MC, Riccardi CJ, Donohue KF, Blagg CO, Christensen RL, Lang AR (2010) A meta-analysis of the effectiveness of placebo manipulations in alcohol-challenge studies. Psychol Addict Behav 24(2):239–253. https://doi.org/10.1037/a0017709
56. Testa M, Fillmore MT, Norris J, Abbey A, Curtin JJ, Leonard KE, Mariano KA, Thomas MC, Nomensen KJ, George WH, VanZile-Tamsen-C, Livingston JA, Saenz C, Buck PO, Zawacki T, Parkhill MR, Jacques AJ, Hayman LW (2006) Understanding alcohol expectancy effects: revisiting the placebo condition. Alcohol Clin Exp Res 30(2):339–348. https://doi.org/10.1111/j.1530-0277.2006.00039.x
57. Bowdring MA, Sayette MA (2021) The effect of alcohol on mood among males drinking with a platonic friend. Alcohol Clin Exp Res 45(10): 2160–2166. https://doi.org/10.1111/acer.14682
58. Bates ME, Mun E, Buckman JF, Vaschillo E, Vaschillo B, Lehrer P, Udo T, Lesnewich LM (2020) Getting to the heart of low sensitivity to alcohol: context moderates low cardiovascular response to alcohol in persons with a family history of alcohol use disorder. Alcohol Clin Exp Res 44(3):589–599. https://doi.org/10.1111/acer.14293
59. Korucuoglu O, Sher KJ, Wood PK, Saults JS, Altamirano L, Miyake A, Bartholow BD (2017) Acute alcohol effects on set-shifting and its moderation by baseline individual differences: a latent variable analysis. Addiction 112(3):442–453. https://doi.org/10.1111/add.13684
60. Conrad M, McNamara P, King A (2012) Alternative substance paradigm: effectiveness of beverage blinding and effects on acute alcohol responses. Exp Clin Psychopharmacol 20(5): 382–389. https://doi.org/10.1037/a0029261
61. National Advisory Council on Alcohol Abuse and Alcoholism (NIAAA) (2005) National Advisory Council on Alcohol Abuse and Alcoholism—recommended council guidelines on ethyl alcohol administration in human experimentation, revised May 2005. NIAAA, Bethesda
62. Schuckit MA, Gold EO (1988) A simultaneous evaluation of multiple markers of ethanol/placebo challenges in sons of alcoholics and controls. Arch Gen Psychiatry 45(3):211. https://doi.org/10.1001/archpsyc.1988.01800270019002

63. Babor TF, Mendelson JH, Greenberg I, Kuehnle J (1978) Experimental analysis of the "happy hour": effects of purchase price on alcohol consumption. Psychopharmacology 58(1):35–41. https://doi.org/10.1007/bf00426787
64. Van Reen E, Tarokh L, Rupp TL, Seifer R, Carskadon MA (2011) Does timing of alcohol administration affect sleep? Sleep 34(2): 195–205. https://doi.org/10.1093/sleep/34.2.195
65. Roehrs T, Zwyghuizen-Doorenbos A, Knox M, Moskowitz H, Roth T (1992) Sedating effects of ethanol and time of drinking. Alcohol Clin Exp Res 16(3):553–557. https://doi.org/10.1111/j.1530-0277.1992.tb01416.x
66. Jones A, Button E, Rose AK, Robinson E, Christiansen P, Di Lemma L, Field M (2016) The ad-libitum alcohol 'taste test': secondary analyses of potential confounds and construct validity. Psychopharmacology 233(5): 917–924. https://doi.org/10.1007/s00213-015-4171-z
67. Meegahapola L, Labhart F, Phan TT, Gatica-Perez D (2021) Examining the social context of alcohol drinking in young adults with smartphone sensing. Proc ACM Interact Mob Wearable Ubiquit Technol 5(3):1–26. https://doi.org/10.1145/3478126
68. Stanesby O, Labhart F, Dietze P, Wright CJC, Kuntsche E (2019) The contexts of heavy drinking: a systematic review of the combinations of context-related factors associated with heavy drinking occasions. PLoS One 14(7): e0218465. https://doi.org/10.1371/journal.pone.0218465
69. Ramirez JJ, Monti PM, Colwill RM (2015) Alcohol-cue exposure effects on craving and attentional bias in underage college-student drinkers. Psychol Addict Behav 29(2): 317–322. https://doi.org/10.1037/adb0000028
70. Waddell JT, Corbin WR, Marohnic SD (2021) Putting things in context: longitudinal relations between drinking contexts, drinking motives, and negative alcohol consequences. Psychol Addict Behav 35(2):148–159. https://doi.org/10.1037/adb0000653
71. Larsen H, Engels RC, Souren PM, Granic I, Overbeek G (2010) Peer influence in a micro-perspective: imitation of alcoholic and non-alcoholic beverages. Addict Behav 35(1): 49–52. https://doi.org/10.1016/j.addbeh.2009.08.002
72. Friedman LM, Furberg CD, DeMets DL (2010) Participant adherence. In: Fundamentals of clinical trials, pp 251–268. https://doi.org/10.1007/978-1-4419-1586-3_14
73. Wang Y, Chen X, Hahn JA, Brumback B, Zhou Z, Miguez MJ, Cook RL (2018) Phosphatidylethanol in comparison to self-reported alcohol consumption among HIV-infected women in a randomized controlled trial of naltrexone for reducing hazardous drinking. Alcohol. Clin Exp Res 42(1):128–134. https://doi.org/10.1111/acer.13540
74. Roese NJ, Jamieson DW (1993) Twenty years of bogus pipeline research: a critical review and meta-analysis. Psychol Bull 114:363–375
75. Fromme K, Stroot E, Kaplan D (1993) Comprehensive effects of alcohol: development and psychometric assessment of a new expectancy questionnaire. Psychol Assess 5:19–26
76. Rueger SY, McNamara PJ, King AC (2009) Expanding the utility of the biphasic alcohol effects scale (BAES) and initial psychometric support for the brief-BAES (B-BAES). Alcohol Clin Exp Res 33(5):916–924
77. Johanson CE, Uhlenhuth EH (1980) Drug preference and mood in humans: diazepam. Psychopharmacology 71(3):269–273. https://doi.org/10.1007/bf00433061
78. Morean ME, de Wit H, King AC, Sofuoglu M, Rueger SY, O'Malley SS (2012) The drug effects questionnaire: psychometric support across three drug types. Psychopharmacology 227(1):177–192. https://doi.org/10.1007/s00213-012-2954-z
79. Haertzen CA, Hickey JE (1987) Addiction research center inventory (ARCI): measurement of euphoria and other drug effects. In: Methods of assessing the reinforcing properties of abused drugs, pp 489–524. https://doi.org/10.1007/978-1-4612-4812-5_24
80. Martin WR, Sloan JW, Sapira JD, Jasinski DR (1971) Physiologic, subjective, and behavioral effects of amphetamine, methamphetamine, ephedrine, phenmetrazine, and methylphenidate in man. Clin Pharmacol Therap 12(2 part1):245–258. https://doi.org/10.1002/cpt1971122part1245
81. Dollinger SJ, Malmquist D (2009) Reliability and validity of single-item self-reports: with special relevance to college students' alcohol use, religiosity, study, and social life. J Gen Psychol 136(3):231–242. https://doi.org/10.3200/genp.136.3.231-242
82. Bohn MJ, Krahn DD, Staehler BA (1995) Development and initial validation of a measure of drinking urges in abstinent alcoholics. Alcohol Clin Exp Res 19(3):600–606. https://doi.org/10.1111/j.1530-0277.1995.tb01554.x
83. MacKillop J (2006) Factor structure of the alcohol urge questionnaire under neutral conditions and during a cue-elicited urge state. Alcohol Clin

Exp Res 30(8):1315–1321. https://doi.org/10.1111/j.1530-0277.2006.00159.x

84. Love A, James D, Willner P (1998) A comparison of two alcohol craving questionnaires. Addiction 93(7):1091–1102. https://doi.org/10.1046/j.1360-0443.1998.937109113.x

85. Leeman RF, Corbin WR, Fromme K (2009) Craving predicts within session drinking behavior following placebo. Personal Individ Differ 46(7):693–698. https://doi.org/10.1016/j.paid.2009.01.024

Chapter 10

Intravenous Alcohol Self-Administration Using a Progressive Ratio Schedule of Reinforcement in Clinical Studies

Elizabeth M. Burnette, Steven J. Nieto, Erica N. Grodin, J. David Jentsch, and Lara A. Ray

Abstract

Operant self-administration paradigms are commonly used in preclinical and clinical studies as a way to experimentally observe individuals' alcohol consumption behaviors, motivation, and/or compulsive use. Self-administration can be assessed using a progressive ratio schedule of reinforcement, in which the amount of effort required to obtain alcohol progressively increases as each reinforcer is earned, is a standard preclinical method used to examine a subject's level of motivation for alcohol. Intravenous alcohol infusion and self-administration in human studies allow for greater control over participants' blood alcohol content and can be assessed using a progressive ratio schedule, creating an opportunity for a cross-species consilience and allowing for the translation of study design and analysis of behavioral outcomes. This chapter covers the implementation of progressive ratio schedules of reinforcement for intravenous alcohol self-administration.

Key words Alcohol administration, Alcohol self-administration, Progressive ratio self-administration, CAIS, Intravenous alcohol

1 Introduction

Alcohol use disorder (AUD) is a highly prevalent and heterogeneous condition, subtypes of which present different characteristics, including individual differences in pharmacokinetic, pharmacodynamic, and subjective responses to alcohol [1–6]. These physical and subjective differences may serve as important risk factors for AUD and may predict the emergence of AUD [7–10]. AUD is characterized by pathological alcohol consumption and failure to limit use despite adverse consequences [11, 12]. Therefore, voluntary self-administration paradigms represent a method through which to experimentally observe individuals' alcohol consumption behaviors, motivation, and/or compulsive use [13, 14].

Melissa A. Cyders (ed.), *Translational Research Methods for Alcohol Use Disorders*, Neuromethods, vol. 201,
https://doi.org/10.1007/978-1-0716-3267-3_10,

1.1 Oral Alcohol Self-Administration

Classic examples featuring oral self-administration paradigms in humans include studies in which participants sample priming doses of various alcoholic (10% ethanol) or non-alcoholic placebo (1% ethanol) beverages and subsequently choose between the sampled beverages [16–19], for which the primary outcome is number of alcoholic beverage chosen or percentage of participants who abstain. Another widely used oral self-administration paradigm instructs participants to consume as many drinks as they like or receive monetary compensation for the drinks not consumed [20–24], for which primary outcomes include number of drinks consumed, observed blood/breath alcohol content (BAC/BrAC) levels, and percentage of participants who abstain. In both of these commonly-used paradigms, disincentives (i.e., placebo drinks; monetary compensation) are included as controls to ensure that choice of alcohol relates to its specific rewarding properties rather than thirst or general reward. Alcohol self-administration can also be assessed with procedures in which animals or human subjects perform a trained operant response (e.g., pressing a lever or button) to elicit the contingent delivery of a volume of alcohol, which acts as a reinforcer to increase the likelihood of that response [15]. Most operant alcohol self-administration paradigms include typical instrumental conditioning factors, such as monetary or effort cost, choice against alternative reinforcement, and various reinforcement schedules. While oral self-administration paradigms benefit from strong ecological validity and relative ease of implementation, these paradigms result in wide variability in BAC/BrAC and low outcome variance resulting from the small number of standard drinks that can safely be offered for consumption in a laboratory setting [25].

1.2 Computerized Alcohol Infusion System (CAIS)

An alternate operant self-administration paradigm is the computerized alcohol infusion system (CAIS), used for intravenous (IV) alcohol self-administration. While this paradigm is limited by lower ecological validity and substantially complex and costly implementation, it presents several advantages over an oral self-administration paradigm, including greater control over BAC, smaller doses of alcohol per administration—which allows for a wider variability of number of "drinks ordered" per session—and automatic safeguards: the response button can be temporarily inactivated at predetermined BAC levels, and IV administration can instantly pause alcohol delivery, causing BAC to immediately decline, which cannot occur with oral administration [25]. Additionally, although ecological validity is a limitation, CAIS also allows for an observation of the direct biological effects of alcohol, as no alcohol cues are present and participants have no previous experience on which to base expectations of the consequence of drinking other than the effects of the ethanol itself [26].

The most commonly used CAIS paradigm allows for free access to a dose of 0.0075 g/dl BAC per administration delivered over 2.5 min, with outcome variables including the number of doses ordered and maximum BAC reached [26–29]. Although free access (also known as fixed-ratio 1 or FR-1) is the most widely-used paradigm, CAIS can also flexibly implement various reinforcement schedules, such as the progressive ratio schedule discussed below.

1.3 Progressive Ratio Self-Administration

The relationship between a response (i.e., button press) and reinforcement (i.e., alcohol dose) is known as the schedule of reinforcement and describes variables, including the number of responses required to produce a reinforcer, the amount of time that must pass before the reinforcer becomes available, and cues signaling reinforcement availability [15, 30]. Reinforcement schedules can then be manipulated, in order to investigate effects on alcohol self-administration and associated outcomes. For example, a progressive ratio schedule may be used to identify a subject's level of motivation to obtain alcohol. Under a progressive ratio schedule, the response requirement (i.e., number of button or lever presses; amount of effort) for the delivery of an alcohol reward increases progressively as a function of reinforcers earned [15]. The last completed level of work (i.e., maximum number of responses) the subject makes in order to receive a dose of alcohol is referred to as the break point and is the measure of motivation for alcohol [31]. Preclinical progressive ratio studies of operant alcohol self-administration indicate that break points are higher in female rodents [32] and when rodents are experiencing withdrawal from alcohol [33]. Additional factors shown in preclinical progressive ratio studies to be involved in motivation for alcohol include sigma and glucocorticoid receptors, conditioned taste aversion and place preference, and genetic interactions [34–37]. Of note, progressive ratio paradigms have also been incorporated into medications development studies to assess the efficacy of novel therapeutics in reducing motivation for alcohol [38–41]. In behavioral economic studies of substance and alcohol use disorders, progressive-ratio schedules are a common way to assess the relative reinforcing value of self-administered substances and can be used to examine the effects of price (i.e., response requirement) on drug consumption (demand curves). Interventions that blunt alcohol's reinforcing value and/or decrease motivation for alcohol in the laboratory may be effective treatments for AUD.

In humans, progressive ratio schedules offer the opportunity for forward translation of behavioral paradigms used in preclinical models and to identify predictors of motivation for alcohol. Clinical studies demonstrate that AUD severity, alcohol craving, and delay-discounting strongly predicted alcohol self-administration, and that self-administration "motivated" and "non-motivated" phenotypes were significantly separable [42, 43]. Additional variables related to break point (i.e., motivation for alcohol) seen in clinical studies include negative urgency (risk-taking impulsivity during negative emotional states) [44], recent drinking history [31], and sex differences [29, 45], which may also be impacted by mood and abstinence duration prior to the study. Progressive ratio studies present an opportunity for a standardized schedule of reinforcement across species, allowing for the translation of study design and analysis of behavioral outcomes.

2 Materials

1. Breathalyzer (Draeger Alcotest 6510).
2. Laptop/Computer.
3. Computer-assisted Alcohol Infusion System (CAIS) [46].
4. Secure dongle for CAIS system and data storage.
5. CAIS monitor.
6. Cable to connect computer to IV pump.
7. VGA cable to connect monitor to computer.
8. Cable to connect IV pump to power.
9. Cable to connect computer to power.
10. USB drive with task information.
11. Response button (Griffin PowerMate NA16029).
12. Infusion Pump (iMed Gemini PC2TX).
13. Infusion Tubing.
14. 20 G IV catheter.
15. Saline solution.
16. Alcohol Infusate Solution (*see* below).

    To prepare a 1 L bag of 6% alcohol solution for infusion (*see* **Note 1**):

Supplies:

(a) 98% dehydrated (ethyl) alcohol for injection.

(b) 1000 mL infusion bag 0.9% sodium chloride.

Prepare under sterile USP 795 standards:

(c) 72.5 mL of ethyl alcohol withdrawn and injected into 1 L saline bag.

(d) Depending on amount of overfill in the bag (prefilled bags generally contain up to 10% overfill volume), the concentration of the final solution will vary between 6.0 and 6.6% alcohol solution.

3 Methods

Our procedure is adapted from an established alcohol infusion protocol [14, 31, 46–49] and includes a progressive ratio schedule of reinforcement. The methods, as published in [7, 42] are described fully below, followed by a step-by-step standard operating procedure.

Participants are seated in a recliner chair with an IV placed in their non-dominant arm. Alcohol is administered using a 6.0–6.6% ethanol solution. Participants are infused at a rate of 0.166 mL/min × body weight in kilograms for men and 0.126 mL/min × body weight for women. The alcohol infusion starts at half target rate, to ensure safety, and is then escalated to the full rate after 5 min of monitoring. BrAC is measured via Breathalyzer every 3–5 min. The CAIS system has been modified to combine two alcohol administration paradigms: a 3-step standard alcohol challenge followed by self-administration. During the challenge, participants receive an experimenter-delivered administration of alcohol designed to reach target BrACs of 20, 40, and 60 mg% (0.020, 0.040, and 0.060 g/dl), each over 15 min. The alcohol challenge results in highly controlled BrAC levels at each assessment (mean BrAC (SD): 0.020 (0.001), 0.040 (0.002) and 0.060 (0.002) g/dl, respectively). Following the 60 mg% time point and a required restroom break, participants begin the self-administration paradigm. Participants are permitted to exert effort (pressing an electronic button) to obtain additional "drinks" through the CAIS system, according to a progressive ratio schedule. Participants are required to order one "drink" to familiarize themselves with the procedure (participants had previously viewed a demonstration). The progressive ratio scale is log-linear and was determined through simulations and pilot testing. Ratio requirements range from 20 responses (first completion) to 3139 responses (20th completion). Each "drink" increases the BrAC by 7.5 mg% over 2.5 min, followed by a descent of −1 mg%/min [46]. A maximum BrAC safety limit is set at 120 mg%. If an infusion would exceed this limit the response button is temporarily inactivated. Except for the first "drink," participants are given no instruction with respect to their self-administration. After 180 min, the infusion ends, the IV line is removed.

3.1 Step-by-Step Methods

1. Initial Measures and Setup.
   - (a) Ensure 0.00 BrAC using a breathalyzer test.
   - (b) Ensure negative urine drug screen and pregnancy test (if female).
   - (c) Confirm age and sex and take vitals, including temperature, blood pressure, heart rate, height, and weight.

   (d) Confirm that participant has been fasting for 2 h prior to infusion session.
   (e) Allow participants who smoke to smoke one cigarette 1 h prior to infusion session.
   (f) Administer any baseline behavioral measures.
2. Set up CAIS computer system.
   (a) Connect pump serial port to computer USB.
   (b) Connect pump cord to back of infusion pump.
   (c) Connect monitor to computer.
   (d) Connect response ("Drink") button and CAIS dongle to computer USB port.
   (e) Ensure that the response button is connected prior to opening CAIS software (*see* **Note 2**).
   (f) Unlock secure CAIS USB dongle.
   (g) Double click CAIS icon.
   (h) Click File > New Session to create a new participant session.
   (i) Select the correct experiment from the "Select Experiment" drop-down menu.
   (j) Input age, sex, height, and weight in the required fields.
3. Set up infusion pump.
   (a) Have nurse insert IV line (20 G IV catheter into a vein in the non-dominant forearm, avoiding AC fossa if possible) (*see* **Note 3**).
   (b) Have nurse take pre-session vitals (heart rate, blood pressure, temperature).
   (c) Plug IV pump into an electrical outlet.
   (d) Connect both IV lines to a single Hep-lock.
   (e) Remove all air bubbles from IV line above pump.
   (f) Ensure that IV lines go through pump fully (particularly air bubble detector on bottom).
   (g) Turn on pump using "Power On" button.
   (h) Ensure there are no bubbles above pump.
   (i) Clear volume infused (Hit "Vol Infused"; "Clear"). Volume infused should be at zero.
   (j) Set pump to computer control using the below steps:
      3.j.1. "Options".
      3.j.2. "4 – Computer Controlled".
      3.j.3. "Enter".
      3.j.4. Ensure that the words "Computer Controlled" are flashing on screen.

4. Alcohol Challenge Session.
   - (a) Have nurse flush catheter with saline and confirm blood flash back.
   - (b) Begin infusion session by clicking the play button on the CAIS program. The words "please proceed with button presses" will come onscreen (*see* **Note 4**).
   - (c) Begin alcohol infusion in CAIS (push button to begin).
   - (d) Instruct participant to press the Drink button when they are prompted to do so (*see* **Note 5**). Instructions: *The alcohol infusion consists of two parts. In the first part, you will be given a standard dose of alcohol. At three different time points during this first part of the experiment, we will be taking your heart rate and blood pressure and asking you to complete some questionnaires on the computer. Throughout this time, we will also be breathalyzing you about every 5 min. As much as you can, please refrain from drinking water during the infusion.*
   - (e) After 3 min, record vitals and BrAC.
   - (f) Continue to record BrAC every 3–5 min until BrAC ~0.020.
     - 4.f.1. Enter each BrAC recording into CAIS after BrAC reaches $\geq$0.010.
     - 4.f.2. Press Ctrl + B or hit the "Enter Brth Value" button.
     - 4.f.3. The "Brth Entry" Window will appear.
     - 4.f.4. Press enter to type in BrAC value exactly as it appears on Breathalyzer (*see* **Note 6**).
   - (g) When BrAC = 0.020 g/dl:
     - 4.g.1. Record BrAC and vitals.
     - 4.g.2. Administer Time 1 Assessment.
     - 4.g.3. Push button to start second slope in CAIS.
     - 4.g.4. Continue to enter BrAC into CAIS every 5 min.
   - (h) When BrAC = 0.040 g/dl:
     - 4.h.1. Record BrAC and vitals.
     - 4.h.2. Administer Time 2 Assessment.
     - 4.h.3. Push button to start third slope in CAIS.
     - 4.h.4. Continue to enter BrAC into CAIS every 5 min.
   - (i) When BrAC $\geq$0.06 g/dl:
     - 4.i.1. Record BrAC and vitals.
     - 4.i.2. Administer Time 3 Assessment.
     - 4.i.3. Use "BB" preload procedure. CAIS will increase infusion for 1 min to compensate for pausing of infusion.

(j) Have nurse disconnect IV lines and have the participant use the restroom.

5. Progressive Ratio Self-Administration Session.
   (a) When participant returns from restroom, re-connect IV lines.
   (b) Press the "resume infusion" button on CAIS.
   (c) Inform the participant that the bar is now open for 2 h.

      Instructions: *We will begin the open-bar part of the study where you will be allowed to use this button to order mini-drinks through the IV. These mini-drinks you order are not equivalent to a drink you would get at a bar, they are less. You will need to press the button several times in order to order a mini-drink. All of the mini-drinks you order will be the same size, regardless of how many times you needed to press the button. At the bottom of the screen, you will see a gray bar. Each time you press the button, the size of the bar will decrease. If you press the button and the gray bar does not decrease, it means the press was not accepted, which may happen if you are pressing too quickly. To get a mini-drink, continue pressing the button until the bar disappears. When the bar disappears, you will begin to receive a drink and you will feel the alcohol entering your system. You may order as many mini-drinks as you would like during the open-bar session.*
   (d) Have the participant order their first "drink."

      Instructions: *The open-bar session has now started. Please press the button now to order your first mini-drink.* [Have participant start pressing the button]. *You can see that the bar decreases each time you press the button until it disappears and the mini-drink is ordered. Please continue pressing now until a mini-drink has been ordered.*
   (e) After participant has ordered their first "drink":

      Instructions: *You have now ordered your first drink. Remember that all of the drinks you order will be the same size, regardless of how many times you needed to press the button. You can press the button for a drink any time you wish. You must continue to press the button several times in order to receive a drink.*
   (f) Continue to record BrAC and vital signs every 15 min.

6. Session End.
   (a) Click the "Data" tab and select "Export" (*see* **Note 7**).
   (b) Remove the catheter and take vitals.
   (c) Press "Volume Infused" button on pump and record the total volume infused.

(d) Provide the subject with a meal.

(e) Collect BrAC every 15 min until participant meets discharge criteria (*see* **Note 8**).

4 Notes

Overall, we have found progressive ratio self-administration to be a well-tolerated paradigm. In our progressive ratio study, only 3 (of 70) eligible participants did not complete the task due to adverse events (two IV infiltration issues and one vasovagal syncope related to IV insertion). Of the 67 who completed the task, no data was lost [42]. This method is also found to have high test-retest reliability and internal consistency [31]. We also observed wide variability in response (*see* Fig. 2 [42]), allowing us to conduct a follow-up study that found clusters of self-administration motivated and non-motivated participants. These clusters were associated with alcohol craving and delay-discounting (impulsivity), such that individuals who craved alcohol more during the alcohol challenge phase and those who discounted future rewards at a steeper rate were more likely to be in the motivated self-administration group [43]. We recommend the differentiation of motivated and non-motivated participants in analysis of progressive ratio self-administration data.

Below are method-specific notes:

1. 6% alcohol infusate solution for administration in humans is unfortunately not commercially available. The clinical research pharmacy will need to purchase concentrated ethanol and saline to make this 6% solution. Please be aware that this increases study cost, as it requires more pharmaceutical support.
2. Make sure the button is always plugged in *before* beginning a CAIS session (if CAIS software has already been opened, close and restart the software).
3. Avoid placing the catheter near a flexion area so movement does not cause the catheter to slip out of the lumen of the vessel. Instruct nurse to verify patency of IV catheter before the infusion starts so as to prevent potential infiltration of ethanol solution into surrounding tissue. Ensure that a transparent dressing is used over the IV site in order to more easily see any ecchymosis or swelling, which may be used as an indication of IV infiltration.
4. We recommend the use of a consistent visual stimulus for all participants during the CAIS session. All participants in our protocol watched "Planet Earth." Pause the video each time questionnaires are administered.

5. Button troubleshooting: If the button stops responding to press, do not stop the session. Unplug the button from its port and reinsert it into a different port on the computer. On the screen, select "View" and "Map Drink Buttons." A pop-up window will appear; select "OK" then press the button again to re-sync.
6. The monitor displays the expected BrAC calculated by the CAIS software. If the BrAC entered is fairly close to the estimate, the box will be green, and you can press enter to confirm the entry. If the BrAC entered is very different than the expectation (e.g., input error, such as entering 0.15 instead of 0.015) then the box will be red, and you can re-enter the breathalyzer reading by pressing R. If not an input error, try taking another breathalyzer reading.
7. CAIS will create a .CAS file of all experimental data, including participant number, date, morphometric data, PBPK parameters, time-stamped BrAC and BrAC trajectory, button pushes, infusion rate, total alcohol used, bathroom breaks, and any comments entered during the experimental session.
8. We chose to have our participants remain in the laboratory for at least 4 h past the self-administration period, regardless of whether or not they chose to self-administer alcohol, so as not to disincentivize self-administration with earlier discharge. Participants were asked to blow into the breathalyzer before discharge to ensure that their BrAC has returned to baseline (0.000 if driving, or < 0.040 if taking the bus or getting a ride). Participants were encouraged to take the bus or get a ride and received extra compensation for doing so.

References

1. King AC, de Wit H, McNamara PJ, Cao D (2011, Apr) Rewarding, stimulant, and sedative alcohol responses and relationship to future binge drinking. Arch Gen Psychiatry 68(4):389–399
2. King AC, McNamara PJ, Hasin DS, Cao D (2014, May 15) Alcohol challenge responses predict future alcohol use disorder symptoms: a 6-year prospective study. Biol Psychiatry 75(10):798–806
3. Newlin DB, Thomson JB (1990, Nov) Alcohol challenge with sons of alcoholics: a critical review and analysis. Psychol Bull 108(3): 383–402
4. Ray LA, Mackillop J, Monti PM (2010, Sept) Subjective responses to alcohol consumption as endophenotypes: advancing behavioral genetics in etiological and treatment models of alcoholism. Subst Use Misuse 45(11):1742–1765
5. Schuckit MA (1984, Sept) Subjective responses to alcohol in sons of alcoholics and control subjects. Arch Gen Psychiatry 41(9):879–884
6. Schuckit MA, Smith TL (1996, Mar) An 8-year follow-up of 450 sons of alcoholic and control subjects. Arch Gen Psychiatry 53(3):202–210
7. Bujarski S, Hutchison KE, Prause N, Ray LA (2017, Jan) Functional significance of subjective response to alcohol across levels of alcohol exposure. Addict Biol 22(1):235–245
8. Bujarski S, Ray LA (2014, Jul 1) Subjective response to alcohol and associated craving in heavy drinkers vs. alcohol dependents: an examination of Koob's allostatic model in humans. Drug Alcohol Depend 140:161–167
9. King AC, Roche DJO, Rueger SY (2011, Oct) Subjective responses to alcohol: a paradigm shift may be brewing. Alcohol Clin Exp Res 35(10):1726–1728

10. Quinn PD, Fromme K (2011, Oct) Subjective response to alcohol challenge: a quantitative review. Alcohol Clin Exp Res 35(10): 1759–1770
11. American Psychiatric Association (2013) Diagnostic and statistical manual of mental disorders [internet], 5th edn. American Psychiatric Association. [Cited 2021 Mar 15]. Available from: http://psychiatryonline.org/doi/book/10.1176/appi.books.9780890425596
12. Grant BF, Goldstein RB, Saha TD, Chou SP, Jung J, Zhang H et al (2015, Aug) Epidemiology of DSM-5 alcohol use disorder: results from the National Epidemiologic Survey on alcohol and related conditions III. JAMA Psychiatry 72(8):757–766
13. Koob GF, Volkow ND (2010, Jan) Neurocircuitry of addiction. Neuropsychopharmacology 35(1):217–238
14. Zimmermann US, O'Connor S, Ramchandani VA (2013) Modeling alcohol self-administration in the human laboratory. Curr Top Behav Neurosci 13:315–353
15. Nieto SJ, Grodin EN, Aguirre CG, Izquierdo A, Ray LA (2021, Sept 29) Translational opportunities in animal and human models to study alcohol use disorder. Transl Psychiatry 11:496
16. de Wit H, Söderpalm AHV, Nikolayev L, Young E (2003, Aug) Effects of acute social stress on alcohol consumption in healthy subjects. Alcohol Clin Exp Res 27(8):1270–1277
17. DeWit H, Pierri J, Johanson CE (1989) Assessing individual differences in ethanol preference using a cumulative dosing procedure. Psychopharmacology 98(1):113–119
18. de Wit H, Svenson J, York A (1999, Sept 1) Non-specific effect of naltrexone on ethanol consumption in social drinkers. Psychopharmacology 146(1):33–41
19. Duka T, Stephens DN, Russell C, Tasker R (1998, Apr) Discriminative stimulus properties of low doses of ethanol in humans. Psychopharmacology 136(4):379–389
20. Anton RF, Drobes DJ, Voronin K, Durazo-Avizu R, Moak D (2004, Apr) Naltrexone effects on alcohol consumption in a clinical laboratory paradigm: temporal effects of drinking. Psychopharmacology 173(1–2):32–40
21. Drobes DJ, Anton RF, Thomas SE, Voronin K (2004, Sept) Effects of naltrexone and nalmefene on subjective response to alcohol among non-treatment-seeking alcoholics and social drinkers. Alcohol Clin Exp Res 28(9): 1362–1370
22. McKee SA, Harrison ELR, O'Malley SS, Krishnan-Sarin S, Shi J, Tetrault JM et al (2009, July 15) Varenicline reduces alcohol self-administration in heavy-drinking smokers. Biol Psychiatry 66(2):185–190
23. McKee SA, O'Malley SS, Shi J, Mase T, Krishnan-Sarin S (2008, Feb) Effect of transdermal nicotine replacement on alcohol responses and alcohol self-administration. Psychopharmacology 196(2):189–200
24. O'Malley SS, Krishnan-Sarin S, Farren C, Sinha R, Kreek MJ (2002, Feb) Naltrexone decreases craving and alcohol self-administration in alcohol-dependent subjects and activates the hypothalamo-pituitary-adrenocortical axis. Psychopharmacology 160(1): 19–29
25. Bujarski S, Ray LA (2016, Nov) Experimental psychopathology paradigms for alcohol use disorders: applications for translational research. Behav Res Ther 86:11–22
26. Stangl BL, Vatsalya V, Zametkin MR, Cooke ME, Plawecki MH, O'Connor S et al (2017, Jan 1) Exposure-response relationships during free-access intravenous alcohol self-administration in nondependent drinkers: influence of alcohol expectancies and impulsivity. Int J Neuropsychopharmacol 20(1):31–39
27. Marcus GM, Dukes JW, Vittinghoff E, Nah G, Badhwar N, Moss JD et al (2021, May 1) A randomized, double-blind, placebo-controlled trial of intravenous alcohol to assess changes in atrial electrophysiology. JACC Clin Electrophysiol 7(5):662–670
28. Cyders MA, Plawecki MH, Whitt ZT, Kosobud AEK, Kareken DA, Zimmermann US et al (2021, Nov) Translating preclinical models of alcohol seeking and consumption into the human laboratory using intravenous alcohol self-administration paradigms. Addict Biol 26(6):e13016
29. Cyders MA, VanderVeen JD, Plawecki M, Millward JB, Hays J, Kareken DA et al (2016, Feb) Gender-specific effects of mood on alcohol-seeking behaviors: preliminary findings using intravenous alcohol self-administration. Alcohol Clin Exp Res 40(2):393–400
30. Panlilio LV, Goldberg SR (2007, Dec) Self-administration of drugs in animals and humans as a model and an investigative tool. Addiction 102(12):1863–1870
31. Stangl BL, Byrd ND, Soundararajan S, Plawecki MH, O'Connor S, Ramchandani VA (2022, Apr 28) The motivation for alcohol reward: predictors of progressive-ratio intravenous alcohol self-administration in humans. J Vis Exp 182

32. Nieto SJ, Kosten TA (2017, June 1) Female Sprague-Dawley rats display greater appetitive and consummatory responses to alcohol. Behav Brain Res 327:155–161
33. Brown G, Jackson A, Stephens DN (1998, Mar) Effects of repeated withdrawal from chronic ethanol on oral self-administration of ethanol on a progressive ratio schedule. Behav Pharmacol 9(2):149–161
34. Sabino V, Cottone P, Zhao Y, Iyer MR, Steardo L, Steardo L et al (2009, May) The sigma-receptor antagonist BD-1063 decreases ethanol intake and reinforcement in animal models of excessive drinking. Neuropsychopharmacology 34(6):1482–1493
35. Sabino V, Cottone P, Zhao Y, Steardo L, Koob GF, Zorrilla EP (2009, Aug) Selective reduction of alcohol drinking in Sardinian alcohol-preferring rats by a sigma-1 receptor antagonist. Psychopharmacology 205(2):327–335
36. Vendruscolo LF, Barbier E, Schlosburg JE, Misra KK, Whitfield TW, Logrip ML et al (2012, May 30) Corticosteroid-dependent plasticity mediates compulsive alcohol drinking in rats. J Neurosci 32(22):7563–7571
37. Green AS, Grahame NJ (2008, Feb) Ethanol drinking in rodents: is free-choice drinking related to the reinforcing effects of ethanol? Alcohol 42(1):1–11
38. Bell RL, Hauser SR, Liang T, Sari Y, Maldonado-Devincci A, Rodd ZA (2017, Aug 1) Rat animal models for screening medications to treat alcohol use disorders. Neuropharmacology 122:201–243
39. Nieto SJ, Quave CB, Kosten TA (2018, Apr) Naltrexone alters alcohol self-administration behaviors and hypothalamic-pituitary-adrenal axis activity in a sex-dependent manner in rats. Pharmacol Biochem Behav 167:50–59
40. Holtyn AF, Weerts EM (2022) GABAB receptors and alcohol use disorders: preclinical studies. Curr Top Behav Neurosci 52:157–194
41. June HL, Gilpin NW (2010, Apr) Operant self-administration models for testing the neuropharmacological basis of ethanol consumption in rats. Curr Protoc Neurosci Chapter 9:Unit 9.12.1–26
42. Bujarski S, Jentsch JD, Roche DJO, Ramchandani VA, Miotto K, Ray LA (2018, Aug) Differences in the subjective and motivational properties of alcohol across alcohol use severity: application of a novel translational human laboratory paradigm. Neuropsychopharmacology 43(9):1891–1899
43. Grodin EN, Montoya AK, Bujarski S, Ray LA (2021, May) Modeling motivation for alcohol in humans using traditional and machine learning approaches. Addict Biol 26(3): e12949
44. VanderVeen JD, Plawecki MH, Millward JB, Hays J, Kareken DA, O'Connor S et al (2016, Aug 1) Negative urgency, mood induction, and alcohol seeking behaviors. Drug Alcohol Depend 165:151–158
45. Plawecki MH, White K, Kosobud AEK, Grahame N, Zimmermann US, Crabb D et al (2018, Oct) Sex differences in motivation to self-administer alcohol after 2 weeks of abstinence in young-adult heavy drinkers. Alcohol Clin Exp Res 42(10):1897–1908
46. Zimmermann US, Mick I, Vitvitskyi V, Plawecki MH, Mann KF, O'Connor S (2008, Jul) Development and pilot validation of computer-assisted self-infusion of ethanol (CASE): a new method to study alcohol self-administration in humans. Alcohol Clin Exp Res 32(7):1321–1328
47. Li TK, Yin SJ, Crabb DW, O'Connor S, Ramchandani VA (2001, Jan) Genetic and environmental influences on alcohol metabolism in humans. Alcohol Clin Exp Res 25(1):136–144
48. Plawecki MH, Han JJ, Doerschuk PC, Ramchandani VA, O'Connor SJ (2008, Dec) Physiologically based pharmacokinetic (PBPK) models for ethanol. IEEE Trans Biomed Eng 55(12):2691–2700
49. Zimmermann US, Mick I, Laucht M, Vitvitskiy V, Plawecki MH, Mann KF et al (2009, Mar) Offspring of parents with an alcohol use disorder prefer higher levels of brain alcohol exposure in experiments involving computer-assisted self-infusion of ethanol (CASE). Psychopharmacology 202(4): 689–697

Index

Melissa A. Cyders (ed.), *Translational Research Methods for Alcohol Use Disorders*, Neuromethods, vol. 201,
https://doi.org/10.1007/978-1-0716-3267-3,

MIX
Papier aus verantwortungsvollen Quellen
Paper from responsible sources
FSC® C105338

FSC
www.fsc.org

If you have any concerns about our products,
you can contact us on
ProductSafety@springernature.com

In case Publisher is established outside the EU,
the EU authorized representative is:
Springer Nature Customer Service Center GmbH
Europaplatz 3, 69115 Heidelberg, Germany

Printed by Libri Plureos GmbH
in Hamburg, Germany